# THE 10 BILLION TREE TSUNAMI

## HOW PAKISTAN IS FIGHTING AGAINST CLIMATE CHANGE AND WHAT THE WORLD CAN LEARN FROM IT

FAHD NAZIR

# ACKNOWLEDGMENT

The completion of this undertaking could not have been possible without the encouragement and support of so many people whose names may not all be enumerated. Their reassurance is sincerely appreciated and gratefully acknowledged. However, I would like to express my deep appreciation and indebtedness to my parents:

Nazir Ahmed Khan and Rehana Tehseen for their endless support and encouragement during this endeavor, without whom this book would not be possible. To all relatives and friends who in one way or another extended their support either morally or otherwise. Thank you!

Above all, thanks to the Great Almighty, the author of knowledge and wisdom, for his countless blessings and love.

Finally, I would like to dedicate this book to all who are genuinely concerned about the environment and climate change. I encourage you to join my Facebook group below and share your ideas and feedback. I would be interested in hearing your feedback and constructive criticism.

https://www.facebook.com/groups/climatechangefahd

Thank you,

Fahd Nazir

CONTENTS

# INTRODUCTION

I worry that I do not worry enough. I am concerned that there is no concern.

There is an unsettling fear circulating our culture that enough is just not enough and that this rollercoaster ride of global environmental change is on its way down, whether we make miracles or not. But that fear is only preceded by doubt and misinformation. The less we understand something, the more dangerous it is; therefore, it is plain to see what needs to be done. Studying, evaluating, and honestly sharing what our impact is doing to the planet, why, and what we can do to reverse it is the only way forward.

It can feel desperate; I understand that. That horrible feeling of not knowing where to take the first step to help yourself and, in turn, others commit to having less of a negative impact on the environment is stressful and overwhelming. We know we all have a part to play in this narrative, but we are not sure what our roles are.

Past predictions were all underestimated. Scientists looked at analysis after analysis, published their findings of global warming, and saw the results in terms of data. However, the rest of the world was still brushing off two world wars and focusing heavily on industrialization and growth. It took time for humanity to be honest with itself, to see through the excuses of development and progress, and acknowledge the fact that it's digging its own grave. And in no way is it more evident than in Pakistan, where the consequences of the country's insecurity have established it as one of the hardest-hit countries by climate change.

Throughout this book, we will ask and answer questions, such as: Why is Pakistan so vulnerable to climate change? What does that all mean in the long run for their people, ecosystems, and the rest of the world? And what have they done about it?

Together we will address those concerns and many many more in the following chapters, where you will have access to the various pockets of information about Pakistan's hot seat, their plans to move forward, and the challenges in their way. This is where you, dear reader, can also assess how far *you* are coming in *your own country* and what you can, in turn, learn from mine. Pakistan has ideas worth sharing, and there should be a lesson for anyone who is willing to listen.

It is obvious, the world is facing a triple threat: climate change, economic crises, and the outcome of the COVID-19 pandemic. Game-changing environmental programs have popped up over the years, but that doesn't necessarily mean that any nation can adopt them. How can a country reeling from past and present pressures overcome its prejudices and unite to change something completely new and foreign in its traditional life? That is where Pakistan is aiming its global responsibility and trying to understand how to upscale their technical know-how as well as their international policies and financial framework with whatever options are available to them.

My name is Fahd, and I care a lot about what I hear, see, and feel. This spurred me to write this book and bring forward the story of my country of birth, Pakistan, even as I sit here in the United States. I have been studying climate change for over five years and have invested much of my time and energy in finding out why, how, and what climate change will do to this planet and its inhabitants.

In my younger years, I followed a very different path than the one I tread now. I graduated from the top-ranked Thunderbird School of Global Management in Phoenix, Arizona, with a Master's in International Business. A bright, hopeful, and successful white-collar job was awaiting me in the Oil and Gas industry. This might seem contradictory to you, but my sheer passion and commitment to the activism of climate change today stem from my past experiences in the energy industry. It took many years through my time spent there before I saw that I was no longer willing to bring to the table the same values and morals as before. I realized how destructive the business could be, and I understood what corporations prioritize in these 'climate agendas' and their immense value on numbers. That is how I arrived at this point in life, asking whether anyone can make a difference and promote actual change outside of political means. Can a developing country like Pakistan transcend the bases of its global status and be the beacon of hope for the rest of the world?

Ebbed on by my intrigue on how Pakistan is carrying the burden and managing the weight of their current situation has inspired me to share. My interest in the policies and government influence that revolves around climate issues was fed even further when I began paying close attention to Pakistan's eco-resilience in terms of community involvement and support.

In the following six chapters, we will be discussing the mentality established by the Pakistan governing bodies to use climate change to their advantage for political and national adaptation and how it is a brilliant move if applied correctly. And looking even further, one begins to recognize a Pakistan that is starting to see a niche, a moment where things can happen and grow. We can call it Pakistan's silver lining in the storm of change.

You will notice that the theme of this book is not about change alone but unity, respect, courage, and most of all, selflessness. It was important to me when writing this book that the reasons for climate change were separated from those of the past capitalistic pretenses, as I understand very well that they still play a prominent role in the bigger picture. But it is vital that the concept of change and modification of behavior come from stronger motifs than power-play and control. I will be putting aside the more obvious reasons for why we are in this mess (human violence, greed, need for superiority) and focus intently on the homogenous actions of a country towards the threats of climate change with respect to the times we live in today. Everyone has different perspectives and predispositions; therefore, my generalized overview is based on research and not necessarily my personal opinion. It is crucial that religion, race, and sexism stay on the fringes of the discussions and that we focus more clearly on the causes and effects of climate change and where these actionable results will lead us.

I will be going through tons of acronyms, policies, and projects, so I encourage you to keep a notepad with you at all times for evaluation and reflection of what you have read. This book is not just a read; please, this book is intended to be a light study of what Pakistan is facing right now and translate that experience to the larger world. Therefore, if you are truly invested, you can assess your own thoughts and motivations toward the subjects put forward by scribbling down what you think and what you can do with your own two hands in your own town, province, or country.

Again, before we begin to crack this subject open, I would really like you to always keep in mind the efforts that your own country is dealing with. My intention for this book is positive, on all accounts, but I would like you to measure this book by your honest values and convictions you have at home, and try to be the judge of whether Pakistan is doing something different, new, or outdated in comparison to your own country. Learning from each other positively, as individuals or entire nations, is the end game.

As was said by the United Kingdom Prime Minister himself, Boris Johnson, this year (*UK PM hails billion tree project*, 2021):

On the subject of planting trees, we're going to do millions in the UK, but I was absolutely blown away. I invite everybody to follow the example of Imran Khan, who pledged to plant 10 billion trees in Pakistan alone, and I think it's very important that we in the developed world recognize our obligation to help less developed countries down this path in all these technologies.

Finally, there is a wonderfully melancholic poem by Sylvia Stults that summarizes everything we are going to discuss in this book. It is equally sad and hopeful, and I believe that it brings a solid entry into this space. The poem begins by describing how our pristine world is being changed by pollution, but then goes on to say:

---

*One can't blame pollution alone*

*As they say, you reap what you've sown*

*So let us plant a better seed*

*Tear out old roots, cultivate, weed*

*Protect what has been given for free*

*Our waters, skies, wildlife and trees*

*For once they're gone, don't you say*

*Consider yourself warned of that fatal day*

— FROM *WARNED* BY SYLVIA STULTS

---

The entire poem, and others like it, can be found on Family Friend Poems, https://www.familyfriendpoems.com/poem/warned.

1

# BEARING THE BRUNT OF CLIMATE CHANGE

*In the 90s, when people talked about climate change, no one understood. But now everyone, especially people from rural areas where they are closer to nature can see the impact of climate change.*

— PRIME MINISTER OF PAKISTAN,
IMRAN KHAN

What does 'bearing the brunt' mean to you? Does it mean keeping quiet and letting the world have its way? Does it mean accepting the unpleasant and hurtful force heaved upon you without concern? No, none of the above. Pakistan is one of those countries currently barely bearing the consequences of its position. To Pakistanis, it means standing up, acknowledging all the external variables, and still making it *their* problem to face the music.

All right, so what do we know so far about the issue of climate change? In a nutshell, it starts with global warming, where the earth's surface heats up due to trapped carbon dioxide ($CO_2$) and other greenhouse gasses (GHG) like methane and nitrous oxide in the atmosphere. The sun's heat entering the earth's atmosphere cannot then leave (due to that blanketed effect), hence the rise in global temperature (global warming) and the devastating consequences it has on the weather (climate change). The fewer of these GHG we add to the natural balance of the climate, the closer we can come to healing.

As you will soon discover, Pakistan understands that they are prepared to address this speed bump head-on and with courageous and innovative plans..

The Islamic Republic of Pakistan is a beautiful land that lies in South Asia, with a varied topography within its borders, making it truly unique and special. To its north lie the highlands, touched by the Himalayas, Karakorum, Hindu Kush, and Suleiman ranges, and all its glorious faces and peaks covered in dwindling forests and ice caps (also known as 'the third pole'). To its east lies the Indus Plain, which holds one of the world's greatest rivers in its basin, the Indus, and its many sisters from smaller peaks. They flow south to the plains and plateaus, where they irrigate semi-arid land that was once a fertile Eden. To the west are the deserts of Balochistan and in the south are the immense coastal mangrove forests in the Sindh province. There are four main provinces within Pakistan: Balochistan to the southwest, Khyber Pakhtunkhwa (KPK) to the north, Punjab to the northeast, and Sindh to the southeast.

Pakistan has traditionally recognized four seasons: a cool, dry winter from December to February, a hot, dry spring from March to May, a rainy summer (southwest monsoon period) from June to September, and the retreating monsoons of autumn from October to November. That was the average estimation in the 20th century, but now we see climate change in various parts of Pakistan affect its people and economy.

The northern regions used to receive the majority of the rain (up to 200 millimeters between July and September), but now we are seeing the monsoons move further and further south and with a more aggressive nature.

The coastal region was fertile in its own right until the sea levels started to rise and the land became saline.

The country was mainly an agricultural-based economy, but now due to the present constraints, it is leaning more and more towards a service-based economy.

While the core regions of Punjab, also called the 'bread-basket' of Pakistan, usually had milder temperatures and fertile conditions for small-scale farming, it is now hurting from flood and drought. Agriculture consisted mainly of cotton, fruits, and vegetables, but during the early 1990s, the agricultural boom occurred and saw more and more high-delta crops—those needing greater water depths to mature, like wheat, rice, sugar cane, and maize—produced mainly for export.

Prime Minister (PM) Imran Khan made a speech at the World Economic Forum last year saying, "Sadly, our governance deteriorated in the last 30 years. That is one of the biggest reasons we have not been able to fill our potential as a country. From now on, my biggest challenge is how we can improve our state institutions, so we can improve our governance, so we can tap our potential" (Harris, 2020).

Because of human settlement and industry factors, we see our climate changing rapidly and without remorse. Later, these patterns will change again and again for the worse, and constant shifting in climate has and will displace billions of people over the next century.

Imran also spoke at the United Nations (UN) General Assembly earlier in September to step in and be the first PM that vows to uplift Pakistan and other developing countries who have been through arduous past political situations (spreading the butter thin). He pleads that developed countries—not feeling the brunt just yet—be made accountable for years of disinterest and disregard in their high emission status, calling it "The scourge of illicit financial flows from developing countries" (*Pakistani Prime Minister Imran Khan speaks,* 2021). And that is the truth of the matter. Plundering the wealth of developing countries and shooting the funds off to their wealthy economies or tax havens is no way to behave.

In this chapter, we will be addressing the situation Pakistan is in, why it is in this situation, and what consequences it is facing from this position.

## A DIFFICULT SPOT

Pakistan has indeed been in a difficult spot over the last half-century. What is being heralded as 'freak events' by climate experts is the constant future of Pakistan unless something happens and fast.

Almost 221 million people populate Pakistan, and over 5% (2 million), according to The World Bank in 2020, are living below the poverty line. Pakistan was rated as the eighth most vulnerable country worldwide by climate change as of July 2021 (Eckstein et al., 2021). This is a staggeringly desperate statistic that cries out for attention and help. While the UN, World Bank, and World Wildlife Fund (WWF), to name a few, have stepped in to fund, support, and regulate Pakistan, it has always remained a high priority for the government of Pakistan that the power to better their land lies in their own hands.

Pakistan is showing great care to push change by paying attention to what the earth is saying and opening up the opportunity for behavior change and capacity resilience, especially in their new government.

The fact that Pakistan is so affected, even though it contributes to less than 1% (Salam, 2018) of the total GHG worldwide, is where the injustice lies. Sure, its geographic placement, just like that of Bangladesh, India, and Indonesia, is not particularly optimal to the climate's reaction to the current $CO_2$ levels in the atmosphere. But that does not condemn a country to its fate, and it certainly does not mean the end.

The agricultural industry has been hit the hardest, as it is the main export and highest gross domestic product (GDP) sector in Pakistan. This change comes from many concepts, beginning with a thorough understanding of the situation, plans to alter their future and increase greener environments, and cleaner procedures and governmental processes.

As reported by the Climate Risk Index, the data received between 1999 and 2018 shows that Pakistan is one of the top five countries impacted by freak climate events. Over 152 significant climate disaster events have wreaked havoc in Pakistan over those years, and Pakistan has lost more than $3.79 billion in purchasing power parity (Recharge Pakistan, 2020).

The most vital concern is the fact that the winter season has almost completely disappeared. Over the last four years, the increase in heavy and erratic rain has caused food insecurity, and without proper water management, water insecurity too. It is vital that communities be aware and concerned about this issue, but unfortunately, they are barely acknowledging it (until recently) as a top priority. Political, economic, and religious affairs took center stage, leaving climate concern secondary.

## WHAT IS HAPPENING IN PAKISTAN?

The question is, who and what are causing this to happen to this country? As expressed earlier, we find that Pakistan is a minuscule or almost invisible (less than 1%) contributor to $CO_2$ levels worldwide, where giants like China at 27% and the United States at 11% (Report: China emissions, 2021) as of 2019 spew the most, with an almost disdain attitude about it.

Many companies are pushing for fewer carbon emissions by 2025, and many large corporations are signing a careful pledge to change their song. But that means little in the larger picture of the immediate transformation, when replacing one statistic for another while hiding devious CO2 emission numbers in other areas is viable.

Let us look at the roller coaster ride that is Pakistan's climate crisis right now.

***Rising Temperatures***

Yearly temperatures are increasing at an alarming rate. Pakistan's post-monsoon winter months were between November and February, but only December and maybe January show cooling in recent years. The season has been getting shorter, and the heat is overbearing in the summer months. Rains are heavy and infrequent, causing trouble for farmers and their crops and displacing communities through heavy floods and crippling droughts. Many now experience an average temperature of 100 degrees Fahrenheit (38 degrees Celsius) throughout the year.

Temperatures in the west of Pakistan are historically the highest: in 2017, the city of Turbat recorded a temperature of 128 degrees Fahrenheit (53.5 degrees Celsius). The heat has three effects: glacial melting, ocean water rise, and extreme heat waves. These three factors are part of just one aspect that surrounds Pakistan.

Many of the measuring stations around Pakistan have not been maintained and kept running, meaning that the fewer stations to measure and record temperatures, the less accuracy in nationwide temperatures and fewer long-term comparisons. Either way, you don't need to see statistics to know that things are getting warmer.

A professor at the Peshawar Agricultural University in the KPK province, Muhammad Akmal, mentioned that "Pakistan's rising temperatures are also being boosted because forests are being destroyed as people clear areas for housing and other development, [resulting] in less fertile land for farming" (Salam, 2018). The temperature rise is in both Pakistan's hands and that of developed countries.

***Drought and Flooding***

The concept that drought and flooding are linked can be confusing until you understand how huge a part the soil plays. Severe summer droughts from extremely high temperatures and little moisture mean that the ground is like a rock and is too dry to hold water. Monsoon season brings unprecedented amounts of rain, where all that water rushes and does not sink, causing flooding. These two very opposing but equally deadly weather conditions are a massive chunk of Pakistan's problem.

We see a drought rate increase in the provinces of Sindh in the south and Balochistan in the west, both arid and semi-arid areas of Pakistan. The hyper-arid predictions are based on the trend of unreliable monsoon rains. This trend is estimated to increase in probability between 25% and 65%.

The majority of the population lives near bodies of water (near rivers and coastal regions) or in the mountains. When glaciers melt, the destruction starts at the top of that mountain and flows through the Indus Basin. The Indus Delta (where all the rivers converge to the Arabian Sea) is affected by both the sea-level rise and the glacial river flooding.

The monsoon season used to be a blessing; now, it is a curse. Something that has been a stable consequence of Pakistan's climate until the last two decades is now turning against them. Farmers and fishermen would set their tasks according to the seasons, but the southerly movement of yearly monsoons, and the aggressiveness, have caused severe issues. The increase of floods in urban and rural areas, which hike deaths and displacement, are also due to unreliable monitoring and damaged infrastructure.

Karachi, in particular, was hit hard in 2020 when they received excess rainfall and flooding over the monsoon season, killing 41 people. The city is below sea level and, its drainage system also failed from inadequate construction and maintenance. They received a record-breaking 345 millimeters of rain in a single day and, the average for the terrible week was around 700 millimeters (*2020 Karachi floods*, 2021).

The irrigated agriculture in Pakistan is not stable and indeed not regulated well enough. Some years too much rain, and some years too little rain, cause huge whiplash effects in communities.

***Extreme Weather Conditions***

The hotter it gets, the more evaporation from bodies of water, which means more potent and destructive storms and weather patterns over the seasonal balance of hot and cold temperatures.

Due to changes in wind patterns from the Bay of Bengal and aggression of the Arabian Sea, the monsoons are moving away from agricultural areas of

Punjab to lower regions of Sindh. This means more rain for coastal regions and less for core regions.

The Arabian Sea has its own global impact on Pakistan. Water temperature is increasing worldwide, and the warmer the ocean gets, the bigger the problem gets (thermal expansion). Melting land-based glaciers in the poles increase the volume even further. According to a 2014 meteorological study (Kanwal et al., 2019), if no adaptations are made for the sea-level rise, then coastal flooding will impact millions of people annually between 2070 and 2100. A higher level of sea salt permeating the soil for miles and miles inland means that high salinity degrades soil and the efforts to plant sustainable vegetation. The El Niño cyclones, which have plagued Pakistan since the 19th century, are another addition to the effects of less rainfall. They shift the rains to other regions, creating drought, and then the following year bring in double the rain.

The danger is the immediate inability of the population to adapt and reconstruct their lives to the new shift in climate. The government must increase its effort to research the new patterns that revolve around living areas, irrigation, and crop patterns and make necessary adjustments.

### *Smog*

The major concern for Pakistan is its integration into a more green and economically stable industry in terms of exports, imports, and local produce.

Smog appears when the nitrogen oxides from car exhausts, factories, and power plants enter the atmosphere and react with sunlight, forming airborne particles at ground level that are highly toxic. Cities like Lahore in the Punjab province are heavily affected during the November and December winter months.

The pollution in these cities especially concerns citizens and the government alike. As massive vehicle use in congested areas and industrial activity radically pollute the air and significantly diminish livelihood. This is also quite visible in the mountainous regions of the Hindu Kush, where large trucks haul tourists up and down those mountains all year round.

Fewer trees equal less distribution of oxygen and removal of $CO_2$ in the atmosphere. The lack of this natural dispersion and cleaning effect from trees is what has pushed Pakistan's country-wide level to increase the flora coverage, thanks to PM Imran Khan. The nationwide 10 Billion Tree initiative needs to work in conjunction with the reduction of low-quality oil and coal imported into the country and the efforts of integrating cleaner industry methods and renewables into the market.

Therefore, you can see that what is actually happening in this beautiful country is purely human and environmental degradation. The air, the water, and the land and animals within are being pummeled on all sides.

## THE EFFECTS

As human nature goes, the causes of the disaster are the second lesson that we learn. The first is the effects. We are all a little late to the party of global warming awareness, but by seeing with our own two eyes what our disregard has done to the earth and the rest of its inhabitants, maybe we will learn to live with the effects and learn how to alter them globally.

The ecology of planet Earth is complicated at first glance, but when you look more attentively, you start to see the simplicity of the reasons for catastrophic changes. These effects start small, degrading land and ice one meter at a time and increasing temperatures in small increments each year. The human eye never saw the change until the water lapped at their feet and the wells ran dry.

Plainly put by the WWF, "More frequent and intense droughts, storms, heatwaves, rising sea levels, melting glaciers, and warming oceans, directly harms animals, destroys the places they live and wreaks havoc on people's livelihoods and communities" (*Effects of climate change,* 2021). Pakistan is seeing dreadful, ill-fated change manifest right in front of its eyes. Rural communities are suffering the hardest, and they have the most to lose. If Pakistanis cannot feed themselves because nothing grows, they also cannot feed their livestock. If there is no more water in the area, then there is no more home. And in some areas, people are standing in lines for up to 8 hours to receive drinking water. This lack of a basic need creates a torrent of poor hygiene, disease, and despair.

With so many climate refugees migrating to cities and higher grounds, Pakistan will lose its diversity in the community placement, and it will degrade further by pure 'groupification' of its desperate people into safer areas. This is never a good thing, especially now after the pandemic. Pakistan is built on its agricultural life, and if the ground grows arid and it is impossible to grow crops or keep livestock, then migration or immigration will become the only option.

A surge in industrialism sparked the need to clear out all the forests and build offices and apartments. Industries boomed, and fossil fuels were all the rage. Now, left in the dust of its own undoing, Pakistan has to sacrifice the health of its people for the upkeep of its economy, which tells you that something is critically off-balance here.

The effects follow deeper and deeper into the past experiences of colonialism and the restabilization of its people after independence. Regaining self-worth, cultural and religious power, and economic scale means that a country's only goal is to survive and find its place in the world again. But Pakistan is now a more mature country that understands its situation, place, and worth. And one of those understandings is evaluating all the effects and what they mean.

***Deaths***

Pakistan has 128,000 deaths per year due to the consequences of climate change. Heatwaves, water stress, flooding, pollution, heavy rain, cyclones, earthquakes, avalanches, and epidemics. Pollution is estimated to decrease the life spans of people in cities between two and five years, and heat waves are increasing (depending on emission pathways) in intensity and frequency with an average of seven waves a year (Junaidi, 2019).

Due to illegal deforestation on mountain faces in summer, heated ice caps, warming oceans, and the government's delay in action, Pakistan is now a country in trouble.

In 2008, children under the age of 15 were particularly vulnerable to diarrheal deaths, and it was estimated that in 2030 around 11.7% of the 48,200 total deaths would be children (Van Bronkhorst & Bhandari, 2021). If climate trends persist, then other climate-related deaths will reach around 17% by 2050, even if diarrheal deaths are projected to decline. It means that if health deteriorates through poor hygiene and poverty, climate change will still become a potent killer through the further knock-on effect by the lack of resources, access, and options in general. This will affect a broader spectrum of people.

Pakistanis are dying from either too much water or too little, no more middle ground. Something has to be done and consequences accounted for.

***Health Risks***

The health and safety departments are the first to need upgrading, as Pakistan's health care stands at 122nd ranking out of 190 other countries by the World Health Organization (WHO). Health care has not been a priority for the government, and their facilities and staff may be undervalued and underpaid. The country certainly does not have a model health care sector and should be financed and supported as much as (or even more than) the environmental initiatives.

Pakistan suffers from a lack of awareness and finance; therefore, you will see many people succumb to illnesses that could be easily treated if the facilities allowed for it. Vector-borne diseases like malaria and dengue fever are big killers. This is due to unsanitary conditions and unclean water. Pulmonary diseases at all ages are common, from general poor health and air quality: illnesses like asthma, acute respiratory infections, and tuberculosis. "Malnutrition, particularly among children under the age of five years, is a serious health issue with nearly half of them estimated to be below their weight for age" (Malik et al., 2012).

And all of this health stress on the shoulders of the population causes further underlying diseases, like mental health, cancer, strokes, and ischemic heart disease.

***Displacement***

Due to the condition of these weather patterns worsening as the years go by, the lives of millions of people who depend on these natural water resources start to dismantle. Apart from the lack of a primal requirement, like drinking water and food (which Western countries take for granted), there is the fact that floods are driving people out of their homes. A prime example is a region in Karachi that has experienced torrential rainfall and enormous amounts of displacement.

This mass climate exodus of people suffering from displacement due to these natural disasters is one of the country's many concerns for its people. Fleeing drought and flooding in the last two decades is a joint event, and there does not seem to be a safe haven for these people to go to. Migrating to the cities has been increasing, and it's like jumping out of the frying pan and into a fire for fleeing refugees. It often gets worse for those refugee communities that dwindle in even more poverty due to not understanding city life and the social divide.

This affects work and schooling in these flooded and dry villages and cities. Submerging infrastructure gets degraded very quickly, especially in the coastal regions, where the salt eats away at everything. The rot left behind when the water recedes sometimes just cannot be fixed and therefore is abandoned.

A report published by Climate Action Network South Asia and ActionAid International stated that if this carries on, by 2050, around 63 million citizens could be displaced (Khan, 2021a). In the rural areas of Sindh and Punjab, people are migrating to the cities, which are already overpopulated. These 'hotspots' of migration need to be studied further and invested so that this form of mitigation can be addressed and supported by the government and health sectors.

### *Food and Water Security*

The matter of food and water security needs to be addressed in a manner that clearly outlines the critical issues and the solutions that can be made by everyone, from the local farmers to government officials.

Agriculture contributes 19,3% of the total GDP in Pakistan. Their 8,5% in the labor force is what sustains 60% of the rural population. As of 2018, 37% of the population is food insecure, and since 1947 the yearly water supply has gone from 5,000 to 1,000 cubic meters per person (Baloch, 2018).

If seasons fail, that means disaster. Farmers needing to be educated is a number one concern. Projects have been established over the years, but the government has not shown much interest in them until recently.

Around 50 years ago, there was a 10% increase in agriculture. Development in cultivation and its modernizations changed Pakistan forever; however, more investment in crops for export over local needs has removed the need for local development of resources. Meaning the country is starving because most of its food and resources are exported instead of consumed within.

Pakistan relies heavily on high-delta crops that require intense irrigation and cropping patterns. Pakistan grows sugar cane, rice, wheat, and maize as their primary export, and these crops will not handle the water shortage the country is facing. The major cotton exports died when the water disappeared, and now they have to import their cotton for their textile industry. Moving to low-delta crops (that also happen to be less $CO_2$ intensive) and innovating cropping patterns remains the task of the government and the flexibility of farmers to run with these new techniques.

Pakistan is sitting with 60% of agricultural land that is going dry because now only 10% of it is being fed by rain. The artificial irrigation of these crops is a further expense, which means that the agriculture industry will be getting the majority of the water resources (that are dwindling, to begin with) instead of the local residents.

Poorly planned and overpopulated is what the Water Board Managing Director, Asad Khan, is pointing at. Yes, these are circumstances that will need immediate attention, but he also mentioned how water transmission in urban areas is under attack, and the law is falling behind. Infrastructure is tampered with, and pipelines are diverted to illegal collection sites and pumped out for exorbitant prices. Officials and police are often paid off by the water mafia and do not even deny these allegations. There does exist a particular police faction that focuses on these thefts and raids and destroys their water pumps regularly, but they tend to just pop up in newer locations around the city and carry on with business as usual. People will go to extreme lengths and pay exuberant amounts for water. It's a big affair!

The terrible tragedy here is not necessarily the theft, but the lack of management of these water supplies. Only one-third of what is available gets used and 70% of that is embezzled out to the richer citizens and the industrial textile sector (some deem this a fake business set-up and that it is actually sent elsewhere). The other two-thirds are lost to evaporation, polluted by seawater, or leaked into the ground through broken pipelines. The demand is so much higher than the supply that it creates gaps for the black market to take hold and manipulate such supply. These 'water thieves' steal borehole water and hijack the Water Board's delivery trucks, selling to the highest bidder at rates of $100 per van load. Karachi is infamous for these dealings, leaving the poorer communities to suffer and wonder what is happening to their livelihoods.

Dams will not fill correctly if there is no maintenance on them and if they are not placed in better catchment sites. And if they are losing more water than is being supplied to local communities and farmlands, then it is an utter waste of construction and maintenance. The accumulated silt (sediment from soil erosion found in dams) deposits in these dams have caused clogged watersheds, blocking essential water flows to other regions.

I believe that the story of humanity will not be told through population increase but population sustainability. We can feed the world more than the world needs; that is a fact. All that needs to happen is the government needs to get a better grip on their conservation of the food and water for 'rainy days' (if you get what I mean). There needs to be a correction plan for distribution, access, and modernized infrastructure.

### *Flora and Fauna and Their Forests*

Forests are declining year by year, and wildlife is becoming scarce. The increased pressure on Pakistan's natural resources and the threat of overpopulation and urbanization threatens those who do not have a voice.

Climate change disturbs the life cycles of plants and animals. With such an increase in nationwide temperatures, plants bloom earlier and survive longer into the colder seasons, while animals' hibernation patterns are disrupted, and their migration routes are destroyed. Breeding season for smaller insects and mammals falters, and food resources change along with it too, where some species will disappear, and others will overpopulate, causing problems. A balance has been broken. Deforestation, extreme weather patterns, severe water stress, and incompatible agricultural practices cause these dwindling numbers.

The International Union for Conservation of Nature (IUCN) is working tirelessly to address the water, forest, marine, and animal species degradation. Both land and marine biodiversity programs are established, but they are not productive, at least not yet. This is all in vain if there is no local participation. It has been noted that Pakistan does not have a very good reputation with animal welfare and many accounts of mistreatment and abandonment are in circulation. There needs to be more accountability of the animal population (domestic and wild) within their borders and education on what rights these creatures have!

### *Coastal Erosion*

Because of many high-populous cities lying below sea level, the threat of coastal encroachment and erosion is evident. The focus is on the coastlines in the Sindh province and other Indus delta regions.

In a journal article from 2019, a study on the erosion of the Indus delta regions over the last three decades showed "that Karachi coastline experiences 2.43 (+–0.43 m-yr) of erosion and 8.34 (+–0.45 m-yr) of accretion, while erosion on the western and eastern sides of the Indus River reached 12.5 (+–0.55 m/yr) and 19.96 (+–0.65 m/yr) on average, respectively" (Kanwal et al., 2019).

Pakistan's coastline to the Arabian Sea stretches for about 990 kilometers, and sea-level rise between Sindh and Balochistan affects important economical environmental assets in these coastal regions. This direct land loss can quickly destroy coastal ecosystems through salinated groundwater, corrosive agricultural areas, and infrastructure destruction. There was a global average increase of sea level between 1993 and 2003 of around 3.1 millimeters. With Pakistan being hit by frequent monsoons and cyclones, this degradation of coastal soil and sea-level rise is accelerated even further.

### *Power*

Pakistan has a capacity of over 30,000 MG of electricity. That is far more than Pakistan needs with its population size. Around 60% of that energy is imported coal, and the oil is of very low quality, Euro-2 to be precise, where other countries acquire a Euro-6. The oil goes straight to the energy and transport sector. The additional 30% is hydropower, a small 6% nuclear, and a measly 4% renewable (Ebrahim, 2020).

The country has five oil refineries and over 20 coal plants in use throughout all provinces, but they are said to be relics for their time, using outdated distribution methods and imported oil that contains high amounts of sulfur and magnesium, adding to the fossil fuel pollution. Pakistan refinery sector policy was implemented in 1997, and since then, no upgrade frameworks for energy production have been seen from within Pakistan's government.

The government is the sole purchaser of power in Pakistan, giving it the monopoly of the market. The capacity charges (having too much electricity but not enough infrastructure to deliver it) of 85,000 rupees (Rs) a year (1 rupee equals about 0.013 dollar in 2021) means that tariffs for electricity surplus are on the shoulders of the consumers and not the government. This year, there were significant blackouts in cities like Lahore, Karachi, and Islamabad, lasting between 10 and 12 hours. The protests only resulted in the government blaming these blackouts on the increased pressure to grid stations and the lack of water in dams.

The government's allegations of malpractice in its power sector fell on the China–Pakistan Economic Corridor profiteering schemes and various detrimental policies. The China–Pakistan Economic Corridor power plants built over the last few years show that power generation is extensively biased and firmly ignores the distribution to the rest of Pakistan.

Due to this circular debt (accumulated debt from government subsidies falling on public distribution companies), which the government has created because of this surplus (a price hike), they have resorted to moving more and more into renewables, like wind, solar, and hydro. While all in the name of recovery, investing more in nuclear to pick up the slack is not always the best idea.

The government's faith in the plans for the cleaner nuclear sector (emission count is close to that of solar and wind) in conjunction with China is said to benefit job creation and power distribution, but the rising temperatures will greatly affect its production. The new nuclear plant in Karachi will use seawater to cool its reactors, but if the temperature of the water is not cold enough, then overheating might occur, not to mention the regular flooding episodes in that area over the last few years.

The Union of Concerned Scientists reported back in 2011 that "during times of extreme heat, nuclear power plants operate less efficiently and are dually under the stress of increased electricity demand from air conditioning use. When cooling systems cannot operate, power plants are forced to shut down or reduce output" (Hersher, 2018). This sector is notoriously dangerous if not operated and maintained super efficiently, seeing many countries, like Germany, shutting down all their reactors. But this clean (and expensive) avenue gives Pakistan hope, along with China's help, to lower electricity costs, lower emissions, elevated distribution, and job creation.

In *The News International,* Tabish Gauhar writes, "Pakistan should always remain a welcoming investment destination, but the key lesson learned is that all counterparties to such long-term contractual relationships must also consider, as a key variable, the financial sustainability, or otherwise, the eco-system" (Gauhar, 2020).

***Economy***

The economy of Pakistan has one goal: keep up with the times. This is critical if they plan on finding innovative ways to adapt financially and socially to the climate change around them.

The Federal Ministry of Climate Change is not performing its critical duties, which weakens its power in influencing domestic resource use, therefore a minimum capacity to advocate for the change. The shrugging of shoulders in government departments like finance and planning cause this lack of accountability and follow-through. "Domestic resources dedicated to climate change have historically stood at a mere 8% of the development budget" (Malik, 2018), which gives you the basic idea of where they have (until recently) stood on the matter.

Inequality has become one of the many byproducts of this disastrous process. Pakistan is fighting for its basic necessities! The government cannot keep up with all the factors that impede adequate water and foodservice delivery: one from mismanagement of infrastructure and prices and the other from the weather. The people fight among themselves, causing tension and stress, and they are disappointed in their leaders for not acting sooner.

The consequence of economic loss is that people are desperate to leave their rural areas and farms, homes, and livelihoods but have to start again from scratch. That is a heavy burden on any government. The more people congregate in cities and urban areas, the higher, more negative aspects will be: more people, fewer jobs, more corruption, pollution, disease, the list goes on. In 2010, around 132,000 square kilometers of land in many provinces were damaged from the ravage of monsoon rains. It is reported that a hefty $4 billion is put into climate change damage annually (Khan, 2021a).

An academic based in Lahore, Farooq Sulehria, expressed that "In the next 30 years, Pakistan needs between $7–$15 billion to tackle this issue. This cannot be done without the support of the international community" (Khan, 2021a). For the economy to come out of this in one piece, the government would need to pick up the pace and figure out how to bring to the table solutions that show a win–win for people and policy. Gearing up for financial stress is part of the agenda, and that makes international support fundamental, especially from the $CO_2$-intensive northern developed countries.

Therefore, the internal architecture of the governance will need to be 'greenified.' They need to bring in politicians who really care about what is happening out there and understand the new and exciting policies and schemes that benefit developing countries through funding and awareness.

So what do you see as the brunt of the situation for Pakistan? Do you think the world should feel pity for this nation, step in without repercussions and help? Or should they leave them to learn and grow on their own terms? The major questions about what is morally and ethically profound about helping often get overlooked by the economic value the country pulls on the global stage. This is only the beginning of their dilemma.

And yes, it is valuable to see Pakistan try to stand on its own two feet and survive the onslaught, but it is not the only country to suffer the consequences of global warming and governing malpractices.

In the next chapter, we look more at why the country is sitting on this terrible conundrum. Whether it is their vulnerability to change or their vulnerability to growing fast enough to adapt to those changes is something we will be getting into with more context and depth.

# 2

# PAKISTAN'S PROBLEM

As in anything in life, we need to first and foremost establish what is in our control and what is not. Once that is understood, we can begin to solve issues by anticipating and reacting to certain disaster scenarios. Figuring out where to adapt to events out of our control, such as climate change, and mitigating manageable and changeable issues, is a problem that many developing countries are desperate to solve.

The impacts of extreme weather can no longer be ignored. The 16th edition of the Climate Risk Index evidently points out that we are at fault in exposure and vulnerability to these events. The main concern is the warning systems and the basic preparedness for those emergency moments that will be more common in the future.

It is always important to note that the financial stability of a country does not mean they are safe from the impacts of climate change. For example, Japan is more acutely aware of its vulnerability in location, emission control, and adaptation and mitigation techniques, even with its modern economy. And the global events of the 2020 pandemic have reiterated and brought to light once again the more systemic and intimately connected factors between risks and vulnerabilities.

Therefore, it is Pakistan's problem to use its strength and build resilience to all the different risks surrounding it, whether geographic, economic, or health-related. Let us jump into their overall contributions to the problem they face with some detail and tact.

## WHY IS PAKISTAN SO VULNERABLE?

We can ask this question for many other countries around the world that face imbalance by similar catastrophes. But with regards to Pakistan, its past, social-economic, and religious facets dictate many of its people's perspectives, and therefore its readiness to adapt and grow.

The list below shows an estimated percentage of what really concerns the average Pakistani right now (WorldEnvironmentDay, 2021):

1. Inflation due to inefficient government policies, 53%
2. Unemployment and poverty, 23%
3. Kashmir tension between India and Pakistan, 8%
4. Water stress and food insecurity, 4%
5. Corruption and bribery within government officials, 4%
6. Disease, 2%
7. National political concerns, 2%
8. Other, 2%
9. Power outages, 1%

One could say that their vulnerability is dependent on their awareness of the issue. If the country does not understand the gravity of the crisis hitting them now, they will suffer from it. They see the effects, but the government and media do not give enough weight to combating the issues through communication and climate education.

The sections below discuss critical issues within Pakistan's geography, economy, society, and expertise.

### *Location*

To begin this conversation, we need to assess the geographical location of Pakistan in the South Asian belt. The specialty of the landscape is a significant contributor to its issues, both environmentally and geopolitically.

Regarding natural disasters, it is contrasting in so many ways, and the impact of climate change on its surface is sporadic, excessive, and instantaneous. Floods affect the main rivers, changing their density, natural flows, and riparian forests (forests that lie near bodies of water) that live on their banks. This destroys crops and kills people and their livestock on an annual basis. "Pakistan's unique landmass which ranges from its high mountain down to the Arabian Sea at a steep incline makes the country extremely diverse and vulnerable to flooding from glacial melt combined with monsoon rain in the summer months" (Webmaster, 2019).

The majority of people live near bodies of water, like the mighty Indus River and the Arabian Sea coastal regions. This significantly amplifies the concern for people's lives during floods and the coastal sea level rising year by year around July and August. What is also evident is the increase in earthquakes near the northern and western areas within the mountains of KPK province and the Hindi Kush area.

In terms of geopolitical issues, its own boundaries do not seem to follow a clear geographic or ethnic guide either. What they do follow is their religious segregation between the majority Muslim and majority Hindu communities. Call it the fault of the British Colonial reign ending in 1947, with a decision to just divide them in such a messy way. Between Pakistan and its neighbors, India and Afghanistan, we see a United Nations recognized territorial disputes in the Kashmir region (India refuses to hold plebiscite per UN resolutions) on the northern India–Pakistan border. From Afghanistan, Pakistan has to face Indian-sponsored terrorist groups wreaking havoc on Pakistan territory.

Their location in terms of sea access makes Karachi the major port for the import and export of products, increasing Pakistan's vulnerability due to minimal alternative options for trade routes. Hence, with the help of China, Pakistan is working towards making the port of Gwadar another alternative port connecting Pakistan, China, and Central Asia.

Therefore, it is easy to estimate why Pakistan is fighting for its natural landscape and its 'intended' borders on all fronts. These conflicts do not give them a good name in the rest of the world, especially if sympathy and assistance are needed.

***Dependency***

Considering that Pakistan relies desperately on the climate-sensitive agriculture department, it just goes to show how the severe shortage of natural resources over the last 30 years has had such a heavy tax on their way of life.

Farming was and still is considered a mainstay for the nation, with four-fifths of the agricultural land being rain-fed and the rest supplied manually. This dependency on the water is often not backed up with enough solid processes to conserve and manage water resources in general.

About 90% of Pakistan's water goes to agriculture, and most farmers will simply flood their fields instead of planning carefully and evaluating resource wastage or adaptive cropping patterns. Most of their crops are high-delta monoculture plantations that affect the overall picture of the water stress, space, and emission from their production. If the farmers were exposed more to cleaner and greener strategies for farming, without it being that costly, then there could be a little less dependency on one source and some self-reliance from each farmer. This is one-fifth of Pakistan's GDP, and if this sector is not coerced to improve its flexibility in crop management, then famine and collapse will soon follow.

Pakistan's dependence on energy through coal and oil is unnegotiable, but it is attempting to move more towards hydropower—with over 30% power coming from hydro dams—and building new ones to keep up with demand.

The problem here is quite simple: price. The cost to adopt a more renewable infrastructure scheme within impoverished countries is enormous! Moving to solar and wind comes at a mile-high price that can often put a country back rather than forward in the grand scheme of things. It sounds great to talk about the encouraging efforts to move towards renewable energy, but the reality is quite different.

***Low Adaptive Capacity***

The major shift in Pakistan's low capability to efficiently adapt to climate change is that the scientific and technological departments are lacking. They are not up to par with the rest of the world's funding and growth in those sectors, making it harder for the country to find solutions within their borders that support scientific research and analysis. This gap of knowledge, in turn, affects forestry, agriculture, health, safety, and fishing.

Another vital point is that of deprivation. There is no general awareness of socio-economic terms, and the country cannot support its healthcare system or rely on it to be prepared for climate disaster scenarios with large influxes of patients. The Pakistani government needs to increase adequate infrastructure and emergency preparedness for these more common occurrences.

With rural livelihoods under attack by climate and other economic factors, more and more people are moving to a service-based form of income in the cities. They are finding it harder to keep up, at times reverting to illegal means to survive. This is a very low adaptive capacity, and education is the only way to open opportunities for change.

If the population cannot receive the correct information and use the right tools, then the words are meaningless and empty. A cause has to be created for the citizens to *want* to change. To become adaptive means to be capable of finding accurate alternative information and using that information with purpose.

Another kind of adaptive shortfall is that of inclusivity. There has to be a more substantial influence on women joining in the ranks and having an input. At the moment, Pakistan comprises 50% women, a considerable force that can be used if given more jobs, more say, and more freedom to help their nation and be a part of the change.

As the proverb goes, "If you give a man a fish, you feed him for a day. If you teach a man to fish, you feed him for a lifetime," so should the government and its citizens think when attempting to reverse low adaptive capacities.

***Lack of Monitoring***

With Pakistan's national health service under immense strain, floods over the last decade have often brought more deaths from feeble recovery efforts and hospitalization than from the impact alone.

There are not enough warning systems that monitor weather conditions and weather fronts. Many countries use emergency preparedness as a way to adapt quickly to the disaster by getting people evacuated and providing safe housing in residential areas and industry. Unfortunately, Pakistan has not created enough strategies and invested enough resources on this intricate and essential need.

The government must address its lack of staffing and general health care facilities in the next few years to be the backbone for the country when there is a certainty that more climate devastation is awaiting in the future.

This country is facing a scenario where they have either shot themselves in the foot for choices made years ago, or they are sitting ducks due to the choices made by stronger nations today.

Pakistan's dependence on traditional lifestyles, along with the unnegotiable terms of government and its neighbors' diplomacy, makes them vulnerable to deviations. Their own inability to cope with this change, mentally and physically, will force them to either migrate or comply. In turn, their downfall in access to higher-quality monitoring systems and health care will only stunt them further in their ability to face these changes.

## PAKISTAN'S CONTRIBUTION TO ITS PROBLEMS

Many like to point towards overpopulation and industry emissions to balance the climate change narrative in poverty-stricken countries. Yes, it cannot be denied that many of the issues facing Pakistan are in its own hands. By not taking action sooner, the already vulnerable geography would only make problems snowball further out of control. The actual self-recognition of where it has fallen short and where it needs to make some significant fixes is in itself incredibly courageous. Speed bumps are bound to appear, but Pakistan is working to keep to the path.

The country's biggest misdemeanors revolve around natural resources (timber, water) and imported fossil fuels (coal, oil, and gas) and their unregulated use in industry. The majority of the populace lacks knowledge on living more sustainably, and the government is not holding enough hands for this to be a complete nation-wide awareness project.

Let us look at the three main culprits with a closer lens.

***Deforestation***

The WWF reported that between 2000 and 2010, there was a loss in forestry of 43,000 hectares every year (*Forestry in Pakistan*, 2019). That means a loss of clean air, stable ground, and ecosystem and all the wildlife that comes with it. The rise in temperature is easily attributed to this mass deforestation. Without enough tree coverage for the soil to hold during heavy rain and floods and no lungs to remove the CO2, means no oxygen-rich air and no heat circulation.

The UN Food and Agriculture Organization has also pointed out that "Decades of felling and natural disasters have drastically reduced Pakistan's forests. Figures for the country's total forests cover a range between 2% and 5% of land area" (Hutt, 2018). This makes Pakistan's disappearing forests one of the lowest in forest coverage, significantly below the 12% safe mark established by the UN.

Forest data scientist Hammad Gilani works at the Institute of Space Technology in the capital of Islamabad. He mentioned that the efforts to plant trees are a great step ahead, but if we do not focus on preventing logging, then it will always be a losing game. "Anybody can go and cut a tree" (Hadid, 2020).

This is especially seen in the mountainous northern KPK province, where illegal logging efforts in conjunction with the melting glaciers from the peaks cause hundreds of deaths due to landslides after flash floods.

The significant problem of a low economy is that the communities and farmers can take their financial burdens out on nature. Going to the local forest and cutting down a tree for the month of cooking and heat, the next and the next, over more extended periods, causes greater damage, accumulating over time. Logging companies do that on a much larger scale, sometimes during the night, and sell the trees to the construction and furniture industries at reduced prices.

The government has initiated laws to stop tree felling, but they were weakly enforced and monitored for many years, often because of corruption, where many of the policies are full of loopholes that truly undermine the value of the strategy.

Then, finally, urban areas were created quickly and without concern for the surrounding environment or their future consequences. An economic growth spurt can do that.

With the unregulated local use, large illegal logging projects, agriculture and livestock placement, and heavy urbanization over the years, the coverage of trees has reduced exponentially all over the country, no longer making Pakistan fertile.

***Reliance on Fossil-Fueled Industry***

As we spoke about earlier, smog is a contribution of various factors, and one of them is the heavy pollution from industry. Brick kilns are used aggressively in the construction industry. They are noted to be one of the world's highest stationary black carbon (particulate matter accumulated through excess Co2, sulfur dioxides, and nitrogen oxides in the air) producers. The 20,000 Brick kilns plants around Pakistan situated in urban areas are responsible for 1.5% of the GDP (Hu et al., 2020) and are extensively uncoordinated and unregulated. This black carbon sits on ice caps and creates atmospheric warming and higher melting rates. This, in turn, affects the water availability in the dryer seasons, when glacial dams break open and flood the lower ranges.

About 24% of total GHG emissions come from fossil fuels used in agriculture and production. Technology has saved us from time and labor but is imposing risks from toxic byproducts. "Sixty percent of [Pakistan's] foreign reserves are used for imports of fossil fuels. The power sector alone consumed 40.4% of total oil used in the country in one recent measure" (Hu et al., 2020).

Coal and natural gas pull a staggering 59.42% of use in generating electricity, and between 2020 and 2021, there has been an increase of 1% (Hu et al., 2020), which is significant in that time frame. This justifies how Pakistan is still under the control of heavy fossil fuels until the day they find a stronger footing (policy and funding) with wind, solar, and hydro.

### *Rural Darkness*

The fact that the world gives us these precious gifts of rain, sun, trees, and its available fertile soil for free is incredible. Unfortunately, these resources will not always be around without understanding and patience, and the world will not always provide them. If this message was more prominent around rural farmland and fishing towns areas, then Pakistan could begin to see some larger mental shift.

Pakistan has the fourth-highest rate of water use globally and does not seem to be stopping anytime soon. The population is increasing, and the demand is high because Pakistanis do not understand how to use water and how *not* to use water. The government is not putting enough effort into distributing that concern towards financially viable solutions of water conscientiousness and storage, and the consideration that there is a need to change the water channels in rural areas and divert them to better grounds.

Overfishing and harvesting created a negative ecological area for the mangrove forests in the Sindh coastal regions, which have steadily been dying away. Human impact through water pollution, soil erosion, and lack of fresh water supply from inland are significant contributors to its decline. When the mangrove disappears, that means less plant coverage for $CO_2$ control and that the Arabian Sea has nothing shielding the coastline from its cyclones and monsoons.

Livestock grazing, overfishing, and cutting down trees for fuel are the practices of the people, as they were for many years prior, without concern or consequence. Now that Pakistanis are starting to see dwindling amounts of everything, they look at the government for help and answers. The answers will not come in simple one-off solutions but a gradual behavior change and more sustainable agricultural and industrial sector implementation in the long term.

The implications for the country are tremendous, and it breaks your heart to imagine a population who have to alter their whole lives because of something they cannot see, touch, or hear. The motion of a country's awareness to the cause is the actual cause! Do you see?

There cannot be a change when 90% of the populace is climate-uneducated and does not entirely understand their situation. Pushing for alternative solutions needs to be communicated to the locals in a context that they can understand and to which they can relate. They would not give a damn if you told them that some country millions of miles away is causing the threat; they would need to know how to address the threat right there, in their land, and what their assistance means to the whole country and, in turn, the world. The story about their future way of life must be told somberly and realistically to push home the goal of reducing and adapting while conserving a nation's traditions, lifestyles, and dreams before it collapses.

If you had to look at the situation from the standpoint of your own country, what would you say the most significant downfalls are? Can you spot the difference in terms of economic value and education? It is vital to me that you, the reader, get to acknowledge and relate in some way to what is happening in a small country somewhere in South Asia.

Taking action in all forms and sizes is where we want to be. We want to stand proud of our nation, no matter the global contribution or lack thereof.

That is the topic of discussion in the next chapter, dear reader. We will be going over how Pakistanis are presently addressing their barriers and what their youth (a substantial portion of the population) and government are doing about it.

Remember, change cannot happen in isolation. If one factor of society needs to change, that gives way for change across the board: something the Pakistan government is oddly aware of, sometimes afraid of, but slowly learning to adjust to.

# 3

# TAKING ACTION

When I speak about action taken by the government and local communities, there are words—like collaboration, pragmatism, support, and education—that bring an overview of the goals set in place.

So far, we have touched on how the nation is being pulled from their homes by the erratic and dangerous weather and that the reliance on natural weather patterns for irrigation is no longer reliable. We have understood that food security comes a close second to water stress, and Pakistan's economy buckles under the overall weight of moving away from fossil fuel dependency to renewables.

The country's population is growing and quite rapidly. The strain on food security will increase when the predictions of an additional 100 million people by 2050 become a reality, not to mention the murderously high level of $CO_2$ present in the urban areas due to the increase in car-clogged roads and intensive industry.

This clogging effect of the human population, heavy low-quality industrialization, mismanagement of infrastructure, and the late arrival of awareness are all forming a conglomerate under the name of insufficiency. To be unable to change certain aspects of the economy because your hands are tied or because the market is not yet viable for this amount of change means that Pakistan has to make some interesting adaptations to both its physical determinants and its physiological determinants. There is a desperate deadline that the government is following closely, and they will need to begin honing in on what they can change *right now*, considering these aggressive implications.

New plans need to be implemented to negate the excess use of fossil fuels and logging within Pakistan. This chapter will focus on how this country is learning to lean into its future challenges and using what it has right in front of it for help. I will be showing you how they are opening their eyes to the renewable sector and how their initiatives and new policies are strategizing better for local and international involvement.

## CHALLENGES AHEAD

According to the Climate Risk Profile of Pakistan, published by The World Bank Group and The Asian Development Bank this year, "An increase in the number of people affected by flooding is projected, with a likely increase of around 5 million people exposed to extreme river floods by 2035–2044, and a potential increase of around 1 million annually exposed to coastal flooding by 2070–2100" (Van Bronkhorst & Bhandari, 2021). Earlier in 2017, the Asian Development Bank reported that without steps taken, the northern regions of Pakistan will be the most severely affected by rising temperatures and decreased water availability in the not-so-far future.

The global average temperature will be superseded by Pakistan alone in the late 21st century, predicted to increase between 34.4 and 40.8 degrees Fahrenheit (1.3 and 4.8 degrees Celsius). Its annual maximums and minimum temperatures will get stronger still and accumulate pressure on an already fragile population. This is undoubtedly certain that rainfall and flooding will increase due to the lack of action on water run-off systems and dam construction. The water resources are heavily undervalued, and the management of these systems needs a reliable improvement..

The water situation in Pakistan is reflected by its neighboring countries. The implementation of the Indus Water Treaty in the 1960s (before global warming became so obvious) was the saving glory for the people of Pakistan and India. The shared trans-boundary water management system worked wonderfully until claims to dams and procedures were put down due to tension between India and Pakistan over the years. The Indus Water Treaty is now facing a tragedy of geopolitical magnitude, where Indian's are interested in staying with adversarial and political agendas instead of working together to combat climate change and water stress in both countries.

An increase in carbon emissions of 300% could be Pakistan's future (considering their growth trend) if they do not align their economic development with sustainable and renewable options. If they plan to become fully functional on hydro, wind, and solar, then they will need to avoid building new coal plants (contracted with China), which is increasing demand on the mining industry and imports. They will have to stop relying on fuel and focus on electric options in transport, production, and general eco-products, supporting both urban and rural communities.

Pakistan would have to change their whole idea about farming and completely integrate greener options to all crops and their yield by reducing chemicals and pesticides and switching to low-delta crops. This would produce a lower output but of higher quality and at a higher price.

And water storage facilities need to run like clockwork, designed with long-term applications in mind and solid quality engineering to allow hydropower to become the support the country needs.

It is crucial that the overall understanding and education of combating global warming happens as a nation. With the projected increase in extreme climate events, risk reduction, planning, and adaptation should take the front stage.

You understand that Pakistan has seen nothing but hardship for a very long time. They are a hard country, a weathered country. They were a reckless country once but have shown a new face to the world and its people. They are formulating a strategy to impose more awareness, involvement, and encouragement to their nation and the world amid an impending climate crisis.

## STRATEGY ON CLIMATE CHANGE

One of the most effective strategies that the country has hammered down is pooling resources and information. Facilitating the assessment of climate risk across Pakistan and organizing research groups to work on this together is a fundamental change in itself.

Pakistan's understanding that each province has its own specific challenges means that the government cannot make a 'one plan fits all' strategy for the entire country. That is why they focus on each condition and institute each solution in a separate and contained manner.

We spoke about Pakistan's own contribution to their problems. We also addressed that they are indeed behind in monitoring awareness, emergency preparedness, and capacity for these more common disasters. And that they cannot rely on their health system to support the implications of climate change, that their behavior is trying to catch up with the sudden need for resilience.

Let us look at what exactly they have been doing to overcome the stress now and what they will be doing in the upcoming years.

### *A National Strategy for Climate Change*

Pakistan had already geared up for the climate battle in 2013 by launching the National Climate Change Policy with a framework to impose a carbon tax, implement reforms for reducing emissions, and move to renewable energy. But now, in 2021, we are hoping to see more involvement in the energy reduction, energy use, and optimization of fuels to support transport.

In early 2019 the new government brought about the Ministry of Climate Change as their 'special unit' in addressing everything related and concerning their environmental impact and what effects those impacts have. One of their main goals is that by 2030, Pakistan will have shifted to 30% electric vehicles and 60% clean, renewable energy.

In June of 1994, the UN Framework Convention on Climate Change (UNFCCC) ratified Pakistan to become one of the first South Asian countries to implement strategies to lessen the effects of climate change on their country and the world. Their goal was to report their inventory on total GHG emissions yearly and initiatives to create sinks (forested areas to retrieve carbon) to remove these sources. From 2003 onward, various landmark developments took place through the Paris Climate Agreement on many workshops, reviews, stakeholder investments, and national company stock-taking. All these plans to group, consolidate, and promote the message of unity in the face of climate disaster are now becoming more urgent.

In 2019, Pakistan issued its Second National Communication on Climate Change, and many of the statistics are hopeful. All these revolve around the mitigation (reduce the severity) and adaptation (changing for better) of the environment, socio-economy, and policy-making when faced with extreme conditions.

Pakistan's Intended Nationally Determined Contributions (INDC) were set in parallel to the expected growth in GDP and have expressed the improvement and adaptation in all health-related departments and education. And of course, there are all the mitigation options in farming, fishing, water conservation, forestry, livestock, urban planning, energy, industry, and transport.

Here are some of the government's most recent steps to support their NDCs:

- Establishing the PM's Committee on Climate Change
- Evident focus on the 10 Billion Tree Tsunami initiative
- Develop the share of renewable energy in the power mix up to 30%

- Integrate the Clean & Green Pakistan and Protected Area Initiative
- Venturing in more solar projects through solar power irrigation systems
- Developing the Electric Vehicle Policy
- Plans to eradicate the use of single-use plastics
- Launch the Mandatory Energy Labeling Regime for domestic products
- Initiate the Ecosystem Restoration Initiative

Through all these advancements, there is increased benefit in creating entrepreneurial spaces and the availability of green jobs throughout the whole nation. Another underlying absolute is the education and training required to allow these points to grow and mature in all the different provinces and their communities. International donors are a huge part of these initiatives to support communities' infrastructure and educational development and government officials towards climate change.

The inclusion and promotion of the youth have been at the top of the agenda, focusing on the digital promotion and new age worldview to spread news and valuable information throughout the country. Clean Green Pakistan and the Green Stimulus (discussed in more detail in Chapter 5) all require young blood to fill the gap of knowledge and interest in their market. The

PMs Special Assistant, Malik Aman, expresses, "Pakistan's youth carry the incredible potential to ensure sustainable development. A resilient youth translates to a resultant nation, and our youth have taken initiatives that carry not only social-economic significance but have an eco-friendliness facet to it as well. The incumbent government is very cognizant of this youth potential and has made it its top priority to engage youth in its developmental interventions" (Vener & Jamshaid, 2021).

We also see the UNs initiative Educate a Girl in Climate as part of the British Institute of Journalism to bring in 1,000 girls in Karachi and educate them into the journalism field. This comes from the lack of accurate information from the journalistic sector in Pakistan and the plan that builds more recognition to promote worldwide vision and technical know-how on climate matters to the larger public.

The amount of funding that has come from within the Pakistani government is honorable, with Rs 47,000 shoved into power subsidies in 2020. "Only two years ago, this amount was closer to 9,000 crore Pakistani rupees," says Power Minister Omar Ayub (Husain, 2021).

We can also see some innovative financing options from the State Bank of Pakistan and the Alternative

Energy Development Board, which intends to allow homeowners to borrow from their mortgage (around $50,000) to pay for solar installations on rooftops in their urban areas. Additionally, the government in KPK province plans to take on 90% of the funding for solar energy in more rural off-grid communities. This, in turn, helps sustain the environment through the lessening of local deforestation for wood fuel.

Civil society organizations and those from the private sector have learned to come together and participate. The enormity of the situation has given everyone reason to stand up and do something about it through participation, support, and most importantly, education. Their coalitions and awareness groups are researching to adapt to Pakistan's environment to the ongoing increase of their climate change.

Educating farmers and laborers on better crop patterns, options available to reduce costs, cutting CO2 emissions from their standard production, or moving to more sustainable practices as a whole is underway. Equally, less water consumption and cleaner crop patterns are required to really implement some adaptation techniques to growing crops: increasing the water management of the areas and learning to store more water in the monsoon seasons while also switching to low-delta crops.

It is not always sufficient for a developing country to implement such high-risk strategies just yet, but the vision is there, and it can come to fruition with the proper funding and policing.

***Focus on Adaptation***

Although mitigating the effects of increased tree cover and farming practices is critical, the importance of adapting with suitable eco-conscious energy options for long-term vision is essential.

Earlier in 2020, the World Bank financed a staggering $500 million to the Resilient Institution for Sustainable Economy (RISE) to assist Pakistan in managing fiscal strain, promote private sector growth under low-carbon restrictions, and include overall transparency. Pakistan needs to build fiscal resilience (creating conditions that support future expenditures) and boost a competitive and robust economy.

The World Bank's lead country economist, Shabih Mohib, said, "RISE supports reforms such as harmonizing sales tax and making the trade tariff structure more competitive. This could help the country attract new investments and spur economic recovery" (World Bank, 2020). This is essential to support efforts to rebuild the country after the COVID-19 pandemic.

Beginning the journey with sound financial understanding and clear budget constraints helps them avoid falling under pressure later. Once these green fiscal reforms are underway, then the aggressive push for renewable energy infrastructure and reliance goes forward.

Wind and solar are abundant resources in Pakistan. Pakistan has a wind corridor in Sindh province of around 110 miles that can receive an average of 7.87 meters per second of flow. Solar is popping up more in cities like Karachi, and there are huge solar panel farms in the Punjab desert where 400,000 panels cover over 200 hectares.

The hydropower potential in Pakistan is 41,722 megawatts, with around 30% reliance on hydro from the country's northern regions. There are over 150 dams at present in Pakistan through the many rivers that run through it.

Therefore, displacing fuel from thermal plants is a great step forward, and this rapid pace of altering the electricity mix will only help. The investment in solar, wind, and hydro are coming forward through policies and international agreements on efficient costs to build, train, and maintain these new establishments.

Water collection points are only now a hot topic of discussion because the availability to collect monsoon and glacial downfall is the only solution. Water catchment zones in higher rainfall areas are already underway in the northern provinces of Pakistan, which still maintains a cooler period and more rain in regularity than the southern regions.

Therefore, one can understand that strategies for emission reduction are at all fronts and work in coordination with one another. Pushing for one edge more than another can bring about an imbalance of the system, so the pursuit of eco-conscious Pakistan needs to be like a marathon, not a sprint.

Adaptation in Pakistan means:

- Beginning the journey to climate-resilient and water-smart communities through increased storage capacities of rain and floodwater catchment sites in dams or boreholes in mountainous areas and the Indus Basin.
- Addressing political oversight in the criminal affairs of water and deforestation techniques. The government needs to be one step ahead of the timber and water mafia!
- The diversification of agricultural practices in cropping patterns and insurance that crops will

maintain, with stronger seeds and stronger breeds of plants and animals.
- Integrating appropriate methods for relocating communities affected by climate change and increasing the monitoring of newer catastrophes to come.

Mitigation in Pakistan means:

- Afforestation and reforestation to create the carbon sinks
- Improved production impacts from energy in industry and moving to renewables
- More effective electrical sources used
- Improved agricultural and processing

The Ministry of Climate Change elevated four critical points:

- The main goal is to ensure the most critical aspects of climate change are addressed and attended to. Implement the short, medium, and long-term goals.
- Each province in Pakistan is committing to take action. They all need to come to an accord, and with PM Imran Khan at the lead, we see a sense of duty to assist and be a name that establishes

sustainability.

- Clean Green Pakistan is the urban forest (urban jungle) project in major cities like Karachi and Islamabad.
- The Protected Area Initiative is focused on the habitat protection of natural areas that sustain many species, expanding the already existing national parks and adding new ones.

So what we see is that Pakistan has big plans; they are gearing up for something really dramatic and sensational, right? But the hype could easily be destroyed with one wrong move from the government, and their image can go down the drain. We will be touching on their shortcomings later on in Chapter 4, but for now, let us look at the various initiatives underway and what they can bring to a country so desperate for good news and so deserving of it.

## THE PARIS CLIMATE ACCORD

At the heart of the Paris Agreement are the INDC. They embody the reduction of emissions and the adaptation of impact from climate change and ratify 197 UNFCCC members under the agreement.

Each country has to outline and thoroughly communicate its actions after the 2020 pandemic. These long-term goals to balance emissions and restoration for GHG sinks and renewable energy use will better the world if implemented with conviction. It is created almost like a challenge for each country.

Recently, I believe Pakistan has stood as a poster child for its contributions in complying and attempting in the best way possible to achieve greater climate goals. Although pledging to reduce carbon emissions by 2030 is not a small bite to take, the way they will do it is eagerly awaited.

It certainly does not help to know that the United States withdrew from the Paris Agreement in 2017 under the guidance of former President Donald Trump. And even though President Joe Biden has vowed to rejoin as soon as possible, a certain mistrust and misguidance on the matter have been growing. Still, the American government might have pulled out from the game, but many American companies are still playing by the rules and doing what they can to lower the numbers.

There certainly was a palpable presence of urgency at the UN General Assembly when the PM urged the 25 heads of state to commit more to their Paris Agreements and strive to fulfill their INDCs at a faster pace.

A few years back, Pakistan could have been considered the laughing stock of the Paris Accord, as they were not showing any credible results to their commitments, with their draft document seriously under-prioritizing their renewable scale. But now, we can begin to see some serious improvement, both in commitment from their behalf and from investment opportunities accredited for smart policymaking and accountability.

The COP26 President, Alok Sharma, mentioned earlier this year, with bleak resentment, that many countries are still not anywhere near abiding by or fulfilling the 2015 Paris Agreement to limit global warming to 2 degrees Celsius. There was no positive news for grand breakthroughs right now. Even though China has pledged to increase the percentage of non-fossil fuel for primary energy consumption by 2030, and India has promised to double its target for renewables, the biggest caution is that many smaller countries, like Pakistan, are finding it hard to meet the Paris Agreement targets by the stipulated date. The PM urges that larger nations fight within their own capacities more vigorously because smaller countries cannot carry the load.

Recently, Pakistan failed to get an invitation to the US Climate Leaders Summit. This did not seem to faze the PM and his government, who saw this slap as an opportunity to create their own summit in Pakistan in the next few years.

There seems to be a growing need for a popularity contest between the major contributing countries and the major impacted countries. It is essential, in recent times, for nations to present the best picture towards their interest in climate change, not necessarily meaning that they will make any changes whatsoever, just that the country presents itself that way. That is why the newer review processes are stricter to each country's INDCs and have increased their range of scientists and experts in various other fields to review and evaluate their findings more acutely. The critical aspect is whether these findings are correct and whether they show more numbers and less actual change. So the value in the Paris Agreement is their strict practice to review fairly, and sometimes harshly, each country's progress.

Moving toward Pakistan's improvement within their INDCs, we see the 'Vision 2025' strategy that shows the government's interest in altering plans and policies for greater eco-targets in all sectors, especially considering their demographic dynamics and increasing energy needs in the future. Addressing their very desperate power shortage and economically valuable friendship with China, there needs to be a wiser consideration in terms of overall fossil fuel reduction and how the China–Pakistan Economic Corridor will help this strategy in the long term.

## CLEAN GREEN PAKISTAN MOVEMENT

A country that sees value in its own people is a good country indeed, and that is the virtue that heralds a new mission for PM Imran Khan. Strengthening institutional trends and policies along with major nationwide behavioral changes is what the focus is on.

The Clean Green Pakistan Movement was born in 2018 and focuses on:

- Safe drinking water
- Sanitation and hygiene
- Solid and liquid waste management
- Tree plantation
- COVID-19 relief

Accelerating the achievement of their predetermined goals is what the government needs. The incentive in job creation substantially increased its value to the communities, especially if these jobs could be pursued long-term.

Volunteers and champions of the movement bring together awareness sessions to propagate the message of climate change to every person they possibly can. Using the power of digital communication in urban areas and eco-religious movements in more rural areas (the context of climate change grows when surrounded by faith) is absolutely beneficial.

Periodic reviews of systems and goal quantification through the Clean Green Pakistan Index are on the government's list by recognizing cities and institutions that have made the largest impact with awards and certificates. Additionally, the Clean Green Champion Program encourages financial and volunteer support from companies and citizens in the challenge.

A partnership with WaterAid Pakistan provides technical knowledge and communication on water management while broadening support through strategies and involvement by private and public organizations.

Again, I would like you to see the trend here in what Pakistan is doing with its policies and projects. By making the fight for climate change competition-based, giving the best prize for most the sustainable and supportive, then the power to want to do better increases as well. Its psychology is rooted deep within the Pakistani culture of encouragement and support and is something that other nations can look to as a guide for their efforts in turn.

As of October this year, the Clean Green Pakistan Movement has supported planting 3 million trees and focused a lot of its media campaigns in schools and universities to promote water, sanitation, and hygiene (known as WASH) programs.

The accountability and responsibility of every citizen in Pakistan during and after the pandemic to uphold the restrictions and health strategies to reduce the spread of COVID-19 is also quite substantial. The distancing, disinfection, quarantine, and wearing face masks have been well handled over the last year or so. Some contribute this to their previous strategies on polio and their very low average population age of 22.8 years.

## RECHARGE PAKISTAN INITIATIVE

The Ministry of Climate Change and the Federal Flood Commission under the Ministry of Water Resources banner collaborated with the WWF to bring the Recharge Pakistan Initiative. They initiated a 30-year vision of resilience to climate change adaptation through flood-risk management unified nature-based solutions to the Indus River Basin. This took actionable steps through:

- Adequate rain harvesting and storage facilities in dams
- Water discharges are regulated through high flood regions
- Smart harvesting of full river flow during the wettest periods of the year

Recharge Pakistan commits to all the other relevant groups, like the National Flood Protection Plan IV and the National Water Policy, all under the Sustainable Developments Goals and the National Climate Change Policy.

It basically states that three phases will be implemented over ten years to revive the water zones across all the provinces. The goal is to increase the overall water storage in floodplains, wetlands, hill-torrents, and the necessary management. They determine which province is more in need and which has more potential for eco-development. They commit to a 2050 goal of partial to mid-partial adaptation to climate change and sustaining water resources in urban and rural areas.

This initiative also works with ecologists in soil preservation on river banks during flood seasons. The riparian forests in some areas along the Indus are precious to ecosystems and are being destroyed by deforestation and the force of floods. Biodiversity, soil, and atmosphere are taking the brunt of it. Microbiomes of the soil and this fragile organic matter that gets tilled tirelessly are now endangered. This is what was termed the 'bleak' episode of soil and is necessary to protect it and let it heal.

Building resilience to climate adaptation is the initiative's number one commitment. They envision that 5% of Pakistan's population (around 10 million people) will benefit directly from this program, both in jobs and in livelihood conservation. The concern is that 80% of river floods are because of glacial melting; the timing of floods and heavy rainfall can be predicted, and catchment zones can be prepared to receive large amounts of water for storage for drought season.

## PROTECTED AREA INITIATIVE

We have spoken about the impact of climate change on Pakistan's flora and fauna. Pakistan needs to remember that with some governance and appreciation, re-establishing a lost natural ecosystem is not impossible.

The Protected Area Initiative grounds the need for "effective governance and management mechanisms for all the 'notified' protected areas (sites that were announced but never protected) in the country," while also bringing to light "the need to increase protected area coverage to ensure the conservation of biodiversity" (Lázaro, 2020).

The initiative has three primary objectives:

1. By 2023, the government, along with the help of the local communities, will need to expand protected area coverage by 15%.
2. Understand the legislative interventions and eco-conscious infrastructure designs when establishing these protected areas.
3. Under the IUCNs Green List of Protected and Conserved Areas, add another seven more national parks.

What we understand about the link between healthy humans and a healthy ecosystem is a newfound realization in Pakistan. So, the launch of 15 new national parks is underway to become fully functional and adequately sanctioned. Pakistan seems to be on the path of overachieving its predetermined goals and is really keen to step further into the boundaries of progress.

At the moment, Pakistan holds:

- 72 sanctuaries
- 66 game reserves
- 19 wetlands
- 9 marine areas

Malik Aman said, "The Protected Area Initiative will prove to be a historic milestone for nature conservation in Pakistan. It will not only inspire and engage our youth in nature-based solutions but also contribute towards global efforts for preserving biodiversity as well as mitigating climate change" (Lázaro, 2020).

Deforestation for agriculture, lumber, and paper production took its toll on the northern regions of Kashmir and Azad Jammu provinces; that is why the job creation of guardians for these newly protected forests and ecosystems is at the heart of it all. This really is a group effort, and through the conservation of forestry, fisheries, wildlife, and eco-tourism, then expectations can be met and addressed with clear guidelines and rules. Around 5,000 jobs will be available to help curve the COVID-19 crash. Through the Green Stimulus (we shall discuss more in Chapter 5) initiative, Pakistan is indeed involving anyone and everyone who is willing to participate.

What we find in Pakistan in terms of forestations are coniferous forests, subtropical dry forests, tropical thorn forests, irrigated plantations, riparian forests, and mangrove wetlands. The conservation of these forests must focus on a comprehensive understanding of the soil, root, plant, and atmosphere. The links in the ecological cycle are very specific, and Pakistan's interest in building a community that knows what these links are, how they affect the environment, farms, and livelihoods is crucial.

I firmly believe that you can judge a country by the way it handles its prisoners and wildlife. We are obviously not talking about the first point at this present time, but the second is becoming a real issue for the government, and I commend them on their newfound interest in conservation and love for their ecosystems.

## GLOBAL EBA FUND

The approach to an Ecosystem-based Adaptation (EbA) is just another way to get the masses thinking and acting. Launched in March of 2021, it is a collaboration between the Federal Environment Ministry of Germany, the IUCN, and the UN Environment Programme (UNEP).

This synergy and co-implementation of various sister companies and foreign investors is necessary to upscale initiatives and bring resilience to the climate change that everyone hopes for. This fund provides seed capital (the finances raised to develop new ideas by individuals and companies with a stake in the project) that helps support catalytic change in climate adaptation and restoration. The resilience of the communities involved is at the heart of the fund. It shows others an urgency to innovate and close gaps in technical knowledge and policy surrounding their environmental perspectives.

Under the EbA fund, we find that many of the previous topics on Recharge Pakistan and the Clean Green Movement are stationed as separate affiliates of the transition. It requires evidence-based identification of sites around Pakistan's four provinces that could be the most effective for complex interventions (adaptive engineering infrastructures). The cost-effectiveness of these projects is vital. The primary goal is to include identified stakeholders willing to maximize positive impacts on the development of ecological, environmental, and socio-economic factors.

Remember, these projects have a subtle commonality: to change the mindset of its citizens to prioritize the green movement and find a reward in seeing the efforts of restoring and adapting lifestyles to conserve and protect what is given for free.

## OTHER ACTIONS TAKEN BY PAKISTAN

So what else is Pakistan doing apart from initiating various projects and funds to assist the communities in bettering their environments and livelihoods?

I am aware that we have spoken about many different projects and programs running in Pakistan and future ventures, but here are some of the more intricate changes that the country is planning or has already implemented for better production, process, and promotion of climate adjustment.

### *Implementing a Cleaner Brick Production*

Implementing a cleaner brick production line means applying an innovative zig-zag method to their furnaces. This means changing the traditional fired-clay brick combustion system that uses tons of coal to fire up the furnace. This current system is costly, both in the emission of black carbon and other pollutants and the increase in coal and fossil fuel use. With this natural draft zig-zag method, we would see the efficiency of brick kiln technologies adapt to the environment's needs. The final products can be hollow bricks or fly ash-clay (use of industrial waste) production, which saves energy, costs about 20% less than traditional manufacturing, and significantly reduces mercury pollution.

The kiln itself would use a circuit that conveys heat in a rectangular formation to zig-zag an airflow path, resulting in better combustion. Instead of using fan drought (using electricity), there is the use of chimney draught (natural pressure) in the kilns. This, in turn, reduces the input of energy resources, reduction in emission count, and safer environment for workers at the kiln.

***Begin to Close all Coal Power Projects***

Pakistan will start the process of ceasing all coal power projects. We have spoken about the country's faith in hydro as their next primary source of generated power for electricity. The PM announced that Pakistan would be closing down two of its relic coal plants costing more to maintain than they produce. This, unfortunately, can mean that the closure is more for production value than for the impact on the environment, but that is still a hugely positive outlook from the government's side.

Although we see the shelving of 2,600 megawatts from coal plants, their replacement in 3,700 megawatts in hydropower is on its way. Hydropower is excellent, but it's not infallible. With new projects in the pipeline at Kohala (1,124 megawatts), Suki Kinari (884 megawatts), and Karot (720 megawatts), this does not mean that there will be no environmental cost to their production. Hydro plants disrupt river ecosystems, produce excess methane in reservoirs, and disrupt sediment flow. This is not necessarily the 'greenest' option at the moment, but it is a good step forward. There will be more reliance on wind and solar in the not-so-distant future, but the appropriate financial mechanics have to be established first, as these are expensive to include and maintain.

It is conducive that the government is focusing on this as a whole, but it is also conducive that they do not pretend that they are doing it for the good of the country and not for the good of their own economy. Instead of throwing attention off what is really important to them (building a stronger and more flexible economy), there should definitely be more transparency between what the government is planning and what the world is told..

### *Alternatives for Rural Communities*

Alternatives for rural communities are what the government is proposing, with stronger feelings of solar and hydro efforts to support these villages and farmers. Harvesting wood for fuelwood is the main practice of locals and communities in rural areas; therefore, providing an alternative is absolute. We spoke about how solar will be the best option for these communities, with projects planned to provide panels and solar cookers for the rural communities, allowing for less consumption of wood and coal.

***Electric and Hybrid Car Conversion***

For some time now, electric and hybrid car conversion is something Pakistan has had at the top of their list. Kalbe Ali states, "The Federal Minister of Industries and Production Makhdoom Khusro Bakhtyar said the new policy would focus on growth and development of the automotive industry together with affordability, quality, availability, and indigenization" (Ali, 2021). This goal is on the ministry and government to increase the tread in reducing oil imports while at the same time looking at electric vehicles as the next viable transport options in their economy. The industry has also suggested hybrid electric vehicles (HEVs) as a more appropriate medium-term solution for government implementation. Due to the estimated cost and outflow of importing the 20,000 units per year, the EVs are unsustainable, so the HEVs would undoubtedly allow the country to still bring in their economic factor (less cost) upon the green movement.

***Liquified Natural Gas***

The country's reliance on liquified natural gas has been around for about six years, but the concern is that the spike in global price within the import structure from northern counties puts Pakistan's reliance under duress. Pakistan's strategy was put together on the premise that these fuels would be cheaper and easier to obtain in the long run, but Asian liquified natural gas rates soared recently and caused a massive backlash from the power industry. This shortfall in governance has put a strain on the government's choice in relying so openly on other countries' spot imports, and their own dwindling domestic suppliers cannot support the demand. However, this can be a win if Pakistan can optimize and push communities to use electricity instead of gas for heating and home use. There is a great need for Pakistan to find an alternative in fuel oil. This also means that its hydropower industry needs to step up and take that gap to its advantage or intervene with eco-electrical products.

### *Agricultural Drip Irrigation*

Agricultural drip irrigation access is what the country is pushing to convert to nationwide. Water-efficient irrigation is another obvious concern, and the arid regions of Pakistan need a certain amount of reliable water. The crops that are produced require a lot of it. Therefore, if the country decided to grow high-value crops instead, like olives and grapes, they will still need plenty of water, but with the new drip irrigation system set in place, it can be used sparingly and at set times, instead of simply flooding the grow area. Striving for a 'more crop per drop' plan, the World Bank, in partnership with Punjab Irrigated-Agriculture Productivity Improvement Project, plans to install 120,000 acres of farmland with the new drip irrigation powered by solar energy. This would cost Punjab's government around Rs 67.5 million ($407,300).

### *Incentivizing Renewables*

Incentivizing the policy for renewables in wind power is the next big step for Pakistan. We can see a potential for growth in the Policy for Development of Renewable Energy for Power Generation, which increases the guaranteed electricity purchase. The policy provides an attractive tariff for inflation and exchange rate fluctuation between dollar and rupee. It allows for the accumulation of carbon credits when a conversion occurs and ensures a zero-duty tax on imported equipment. These are just some of the advantages of sustainable development and the benefits that incur.

## WHAT THIS MEANS

Overall, what we see from all of these incredibly innovative and powerful models for development is that they can rocket Pakistan's own economy forward, but only if they have the courage to take the risks necessary to make these renewable options available and wait for the time when they become profitable.

I am sure you understand that by taking eco-conscious risks, the possibility of a loss is there (nature is unpredictable). However, a failure to not act and follow through on its commitment is even a greater loss. If Pakistan is willing to wait for its renewable change to prove financially viable, then we could see some really wonderful things come out of the country. Suppose Pakistan is cautious and content with its current level of growth without really perceiving the long-term changes to come. In that case, remaining fossil fuel-dependent and staying within the good graces of the giants, like the USA, it's not going to change any time soon.

So, what do you think is going on in your country? Are they making any progress towards changing the perspective of climate change through the media? Are you seeing some existing programs and initiatives that positively impact the way communities consume and waste, respectively? Does your country show that it cares?

Do look for exciting proposals on your own country's use of energy, imports, export power, or overall domestic production. Take careful note of where the biggest concern for impact is? And around those factors, is your government actually implementing positive procedures? Most conversations and topics these days have a short lifespan before disappearing from our thoughts. However, discussion on climate change will remain with us for many years in the future. Therefore, it would be good to know the various tidbits of information covering your own country's fight towards overall CO2 reduction.

So let us jump right into the next chapter and focus on the subject that probably piqued your attention to this book in the first place. I want to take you through how a gigantic project, the 10 Billion Tree Tsunami, became the flagship for a government's growth and gained global recognition.

# 4

# FROM 1 BILLION TO 10 BILLION TREES

It started with fear, then it became understanding, and with love and support, it changed.

A project this size was always commendable. The hope that something as humble as planting a tree could impact Pakistan's future was the message that spread across the nation and the rest of the world. And this message was a challenge for the rest to take it on themselves and look within their own borders for ways to mitigate the effects.

As you may know, trees are the lungs of our planet and retrieve $CO_2$ from the atmosphere and convert it into oxygen. The more forest coverage, the more carbon sequestration (retrieval and storage of $CO_2$ from the atmosphere) can occur in these newly integrated carbon sinks (forests).

There are two very clean and accessible ways to combat deforestation: *afforestation* (creating a forest from scratch with saplings) and *reforestation* (extending certain areas of natural damaged forest areas). By creating a new forest, expanding established ones, and protecting them fiercely, the $CO_2$ count above Pakistan's land will decrease and stabilize, but it will take time. These trees are only one niche of the bigger solution to climate change and work in conjunction with many other policies towards a greener Pakistan.

The government approached this challenge in the standard way:

- **Vision:** what they hoped to achieve in the medium to long-term.
- **Target:** what climate facts did they want to see reported through this change.
- **Strategy:** what they were doing to wrestle with the problem and who helps.
- **Project:** putting it all together and running it with precision.

Job creation was inevitable and welcomed with a project of this magnitude, where restoration of community values and 'back to your roots' ideologies resurfaced. Through these challenges, a realization began to settle in: that we are in this together.

And together meant the whole nation! Men, women, children, and youth are pulling their forces together, under the guidance of their government and officials, to prepare their landscape and allow future generations actually to have a future. This incredible effort of mitigating the effects of climate change is positive, but we will also be looking at whether they might have been generalized and overestimated, both in costs and in benefits.

## THE BILLION TREE TSUNAMI

It all started in 2015, when Imran Khan, former cricketer and Chairman of Pakistan Tehreek-e-Insaf (PTI) political party, was thinking big. He began looking closely at his governed northern mountainous province of KPK, where 40% of Pakistan forests lie, and a majority of deforestation occurs.

The Green Agenda was established and rode its success with the cracking down on illegal deforestation in KPK province between 2013 and 2017. During those five years, the PTI increased forest coverage by 22% in pine trees along the mountainous landscape. This was a success, and Imran looked at his Climate Advisor Malik Amin and suggested increasing that number to a billion and calling it *The Billion Tree Tsunami Program* (TBTTP)!

The PTI then took it upon themselves to address this issue within the 2017 Bonn Challenge (global restoration of 150 million hectares by 2020 and 350 million hectares of the landscape by 2030) by implementing the Billion Tree Tsunami Project. Pakistan accomplished its initial pledge to reach 150 million in 2020 and surpassed it with a whopping 350 million trees planted, making Pakistan the first country to complete the targeted amount. This could only put the wind at Pakistan's back to go further and show the world how diverse and driven they are.

This project would reinforce riparian embankments (root systems on the retaining walls in river systems) in catchment spots along the province's Indus, Swat, and Kunhar rivers. Trees have been planted in agricultural settings as well to improve the biodiversity alignment with livestock and crops.

"This project is naturally restoring a previously deforested landscape, which will assist meeting present and future needs and offers multiple benefits for climate adaptation and mitigation in a very climate-vulnerable province [KPK]" (*Pakistan's Billion Tree Tsunami*, 2017), says Project Director of the Billion Tree Tsunami, Muhammad Tehmasip.

The main requirements for this project are flexibility, transparency, and community involvement. By bringing in stakeholders, forestry sectors, and communities, large involvements and collaborations were established.

Here are the main elements that were considered determinants to the success of this project:

- *Pushing back illegal deforestation*: Sure, there would have always been an issue with the logging companies who were often supplied by the 'timber mafia' (illegal logging groups), causing a loss of supply at lower costs due to the initiative of taking back the forests. This was a hard pill to swallow, so KPK province had to start importing wood for local use to combat the bigger issues. This was the first time Pakistan saw resistance to the lawlessness through collaboration between government entities and the public, and more than 600 illegal sawmills were closed and destroyed. Their zero-tolerance was the catalyst for the program to become more pronounced and included the local communities cutting down trees for fuel (their only energy source). This was a little harder to get around as rural communities could not be monitored; they needed to be taught why cutting trees was detrimental and given other options for cooking and keeping warm. The government has been discussing supplying eco ovens and gas cookers to local residents, but this again requires the proper funding, the right plan, and requests.

- *Establishing transparency*: The clarity and credibility of the project had to be visible to the world and the citizens. Third-party verification was necessary (another first time for Pakistan) by the WWF to monitor and evaluate their progress. The constructive criticism and praise through yearly reports were really beneficial for the project to perform on par or above. Learning where they are doing some good and where they could pick up the slack was what gave them a solid guideline and motivation for future projects to come. In conjunction with performance auditing, the PTI created a website that could communicate and share its statistics and overall message to the public. This improved the collaborative effects and grew its popularity.

- *Strongly involving the community*: The youth was the program's primary focus, as were women in their communities—the project created around 500,000 green jobs for local residents in the protection and planting of trees. A 60% of the labor went into assisted natural regeneration and the other 40% into afforestation. Around 13,000 nurseries were established to grow the saplings before plantation, which employed youth and women to learn more while actively helping the project and boosting the local economy. People were stationed as *nigehbaan*, or guardians, to protect the already vulnerable areas from further harm. These particular forests needed more time to achieve self-restoration on their own; therefore, they needed to be guarded: hence, assisted natural regeneration. This and planting new trees in other areas brought about a significant behavior change, thinking, and new green employment skills.

The support from the PTI's own pockets towards the KPK province restoration was remarkable, and they issued US$ 120 million to jump-start the project and a further US$ 100 million to carry it through to 2020. This was the first time such a large amount was allocated to ecological issues. The WWF's recent audit of the TBTTP concluded that they had managed to plant over 872.3 million saplings with a positive 89% survival rate.

## THE 10 BILLION TREE TSUNAMI PROGRAM

Thanks to the immense popularity and media coverage of PTI's investment in its Billion Tree Project, Imran Khan became Prime Minister in 2018, after many years of power in KPK province alone. Their slogan of *Tabdeeli Arahi Hai,* meaning "change is coming," was quite appropriate for their next step of action when the Green Agenda went national.

One of the first things they did once in power was to initiate policies and actions that were to be implemented over the medium to long-term goals within more provinces. If they could achieve a billion trees in one province, what could they do with the whole of Pakistan? Therefore, the gigantic and noteworthy 10 Billion Tree Tsunami Program (or 'Plant for Pakistan') began in 2019 and is projected to meet the goal by 2023.

The Federal Minister of Climate Change and Special Advisor for PM Khan, Malik Amin Aslam, made a statement at the 2021 World Environment Day, saying,

"Our government really feels that this is something we have to do for this generation and the next generations and the ones who come after that." He also mentioned, "We took this option of planting trees because we believe in not fighting nature, but in making nature an ally to solve the issue that we have with nature" (Hess, 2021).

Yes, it used many of the already established effectors, goals, and objectives that were in place for the Billion Tree Program in the KPK province, but now they had to multiply that by countrywide numbers and include other political dividends. That meant ramping up all those previous collaborations from earlier projects, bringing in more international investors, and powering up the citizens of Pakistan to see exactly what could be done.

The help of social media and the influence that PTI used on the youth to demonstrate their message and vision got them where they are today. Not many other political parties were reaching out so far and so openly to the public, focusing their efforts on the environment rather than geopolitical and economic factors (at face value). Their political dividend was inspired, which is one of the primary reasons they won power in 2018.

Not only do we see this immense project reach out for climate equity and change, but it is creating thousands of jobs. Almost 85,000 people were employed in planting, and 700 jobs were created in protecting vulnerable areas.

Already we are seeing the first phase of the project at hand, with 3.5 billion trees projected with a cost of Rs 105 billion. In 2019, some 1.6 million different species were planted, like the eucalyptus (10% to 12%) used in waterlogged areas, and others like peepul, spruce, pine, and neem planted around the country in various cities.

Grafting and planting saplings of olive trees in the provinces of KPK and Balochistan is a nifty new move because olive trees hold a tremendous amount of water and are suited to the current climate and soil. Creating the 'Olive Valley' in the Potohar region has potential for remarkable economic growth in olive oil exports (Pak Olive Company). They plan to plant over 10 million wild olive trees in KPK province and neighboring areas in 2026.

Mangrove wetlands have focused around the coastal regions, and stopping the encroachment of coastal waters is another large project. Pakistan has one of the largest mangrove forests in the world at its coastline, and these are the natural barriers to coastal storms and rising sea levels.

This project also benefits the wildlife of Pakistan, fostering them to return in numbers and help restore the imbalance that has been created: mammals like snow leopards, Himalayan brown bears, caracals, chitals (deer), and various birds and insect life.

## PRAISE FOR THE PROGRAM

There has been a worldwide acknowledgment of Pakistan's efforts over the last five years. From international developed countries like the United Kingdom and Germany, Pakistani ex-patriots worldwide, climate activists, and even celebrities have shown an appreciation for the grandeur and the courage.

UNEP Regional Director, Dechen Tsering, said recently, "Large scale restoration initiatives such as the TBTTP are central to Pakistan's efforts to support the UN Decade Ecosystem Restoration program." He also added, "We are at a point in history where we need to act, and Pakistan is leading on this important effort" (*Pakistan's Ten Billion Tree Tsunami*, 2021).

Environmentalist and graduate from the German University of Heidelberg, Sitara Parveen, used to live in the Hunza region in northern Pakistan. She says, "I have seen the planting of thousands of trees in my area," and "people are being encouraged to plant trees which would help prevent natural disasters as well as providing a healthy environment" (Khan, 2021b).

UK Minister Lord Goldsmith made a statement in Parliament recently, saying, "The project is one of the most ambitious tree planting initiatives in the world and is a successful precedent for others to follow. I absolutely enthusiastically commend and celebrate Pakistan's Ten Billion Tree Tsunami and the tens of thousands of jobs that have been created due to the project. It goes to show what is possible and what can be achieved" (Dawn Staff, 2021).

Director at the IUCN, Mahmood Akhtar Cheema, says, "Pakistan is emerging on the world map for its efforts for projection of its natural capital. This event will further bolster its position and provide a chance to convert the simmering environmental challenges into opportunities" (Latif, 2021).

Pakistani Member of the House of Lords in the United Kingdom and board member for the WWF Pakistan, Aamer Sarfraz, said, "Pakistan is showing global leadership in tackling climate change. The project is an effort that all Pakistanis can be proud of. There is no doubt that 2021 is the year of planet Earth, and by the 26th session of the UN Climate Change Conference of the Parties (COP26) due to be held later this year, we hope to celebrate more success from Pakistan" (Dawn Staff, 2021).

At the 76th UN General Assembly, UK Prime Minister Boris Johnson hailed Pakistan's achievement by saying, "We are going to plant millions of trees in the UK, but I was blown away [by the Billion Tree Project]. I would like to invite everybody to follow the example of Imran Khan, who has pledged to plant 10 million trees in Pakistan alone" (*UK PM hails Billion Tree Project,* 2021).

And on their social media page, United Kingdom rock band Coldplay pointed out how unique the idea was, posting the link on their feed, which was an incredible media boom thanks to their millions of followers. The more people talk about it, the more it becomes a reality. The world will not forget the forest as long as communication and support are propagated within and without borders.

## MONITORING

We have spoken earlier about the project's need for transparency, and monitoring is vital to create factful, honest, and transparent reviews of performance. Collecting, analyzing, and using that information constructively is the fundamental priority. But what really sticks is the action of sending it back to the government to pore over and decide what is working and what isn't, always keeping their standards clear.

The unbiased and effective monitoring from WWF Pakistan started during TBTTP back in 2016, where the first report was sent covering Phase 1 activities. They would monitor KPK province in its watershed management, dam maintenance, afforestation, and farming practices. The team consisted of 18 professionals from environmentalist groups, foresters, scientists, and experts that went around the province to sample and record what they found.

In accordance with the Ministry of Climate Change, WWF, and IUCN, field data has been collected periodically in forest coverage and through remote sensing all around the country after 2018. Technical expertise will support the watershed management and landscape restoration, along with the plating and protecting the restored forests. The UN Sustainable Development Goals are upheld and targeted, allowing the government to enact changes if need be and bump up financial backing from foreign investors.

Feedback on progress is and will always be the fundamental premise for the whole plan. If there is no accountability for the failures or mistakes made by the governing bodies and their policies, then there will be no lessons learned. By honestly reviewing the practices set forward (especially for such a young country just beginning to understand its place), there can be an open space for real growth, development, and thorough momentum toward climate change.

## CRITICISM

People will always have something to say about grand proposals, projects, and plans. The negative comments have come in from far and wide, criticizing the short-sightedness and un-taught nature of the scheme. The criticism ranges in political, environmental, and social disagreements on how the project is stationed, why it is really there, and who it really benefits.

Some believe that the afforestation of some areas of Pakistan will affect farmlands and economic cycles. Because grazing animals are great preventers of wild-fires in extremely dry and hot temperatures, removing them for plantations could be a dangerous maneuver.

Others, like Hassan Abbas, an environmental expert, say, "It is not experts who are in charge of this project but bureaucrats who know less about deforestation and causes of deforestation. They do not even know these terminologies, let alone understand the real causes that are destroying the environment" (Khan, 2021b). Without the proper empirical planning, it can be more destructive than helpful due to the severe water stress and irrigation mismanagement.

Skeptics like Ali Tauqueer Sheikh, who is a climate expert and CEO of LEAD Pakistan, cries about the injustice of the nature performance bonds by saying, "It is about money and not about biodiversity, climate change, emissions or carbon sequestration" (Zaman, 2021). He points out that this scheme to deflect pressure from global policies and renew an image of piety to the world is its only mission, no more, no less. That puts a damper on the ideals and suggests that these many projects out of Pakistan's government are not what they seem.

The concept of transparency is also on the chopping block for Pakistan as Inspector General of Forests, Syed Mahmood Nasir says, "Imagine billions of trees and calculate the total cost, which could have been diverted for cleaning the river banks, ensuring the flow of water from rivers that would have led to the natural growth of trees without any money" (Khan, 2021b). It is estimated that one sapling requires around Rs 15 of water, and the new plantations can suck dry natural water springs that local communities need for survival. The water price is great, and the frustration this early in the project is natural, as people want to see change now, not in years' time.

Of course, the political aspect of this power struggle was at center stage, too. Some claimed that the nurseries set up for sapling growth were in the hands of PTI supporters and that the government buying from them directly is biased and unconstitutional. Even the report of 85,000 jobs created through TBTTP is disputed by officials, some saying that it has tampered with data without evidential proof of employment.

The National Accountability Bureau in January 2020 reported that a total loss of Rs 462 million ($3 million) was found on the budget scheme for the project. Intense allegations of corruption through embezzlement, misappropriation, and 'ghost labor' were said to be the cause for the loss of public funds.

Pakistan is rated 124th in the 180 countries that are part of the Transparency International Corruption Perceptions Index (Zaman, 2021). The concern that the funds are not going where they are truly needed is of the highest attention. "Pakistan has allocated Rs 1.37 trillion for defense services in the 2021–2022 fiscal year, a 6.2 increase" (Zaman, 2021). That means that even if they are preaching focus on ecosystem intervention, they are still prioritizing political agendas between neighboring countries and the pressure does not seem to be going away.

Other critical assessments of the scheme are:

- The project is too long-term. Trees take years and years to grow big enough to absorb enough carbon and be considered real $CO_2$ sinks.
- Insufficient and inefficient metrics to measure the amount of carbon the forests can absorb.
- What kind of planting techniques are critical; the type of forest must not be monoculture-related, as that can negatively impact biodiversity.
- Human displacement is not only by weather events but by the government making space for their high-delta plantations.
- Executing this plan far too quickly and without adequate thought harms people and environments.

## ADDRESSING THE CRITICISM

In a country that has dealt with violence throughout most of their independence, with patriarchal refusal to change to more modern worldviews, a little criticism in the face of change is not that bad. The political, religious, and social backlash was anticipated and hopefully prepared for.

Ali Tauqeer Sheikh, the founder of the environmentalist group LEAD Pakistan, openly stated, "We have been consumed for so long by so many other challenges such as the war on terror that has engulfed our cities, suicide bombings, public health, that kind of thing," adding, "It is extremely important also that we ensure we have enough fresh water and that our development does not destroy our own coastline. We have one of the largest deltas in the world, but it is dying because of climate change" (Jamieson, 2018).

He mentioned that even with all of the past experiences haunting Pakistan, the awareness of the projects is the number one concern. To make the population understand and let the world see that adding lungs to the earth is what will support their environment one tree at a time, not necessarily save the world! The citizens are holding the PM accountable for his choices and are holding their breath (not too long, I hope) for clean air to return to Pakistan.

It is noteworthy to say that the processes that the government went through to begin the project was grounded and trustworthy. They had all the proper governing parties, affiliates, and communities on board, all ready and committed. Criticizing before any real change is seen can feel like a nit-picking agenda with aggravated past pressures dictating mistrust. But this is natural, and we can only wait and see what will come out of it with a positive and pragmatic mindset.

Addressing the criticism requires backing from hard evidence of actual change and community support. The PM advocated for the Japanese Miyawaki method that grows trees 10 times faster and 30 times denser than other reforestation methods. This replication establishes the plantation of dozens of indigenous species back into their original ecologies in one same space. The PM has used this method in various cities around Pakistan (Urban Forest Project), and it is best seen in the Jilani Park in Lahore, where it turned from arid to lush.

This dense native environment brings back wildlife and insect populations into that area and works as an effective carbon sink. It is very important that the soil surveys are accurate, irrigation is effective in the local climate, and maintenance of these areas is monitored regularly. Expertly planting the three layers (sub-tree layer, tree layer, and canopy layer) of shrub diversity for increased growth also requires no extra fertilizer and insecticides. If monitored correctly for a full three years, it will eventually become self-sufficient and regulate on its own.

As they say, Rome wasn't built in a day. You cannot expect a country that has struggled with typical developmental issues like corruption, violence, and poverty to jump out of it with a little media attention and support. As I have said before, the real change comes from the people and their outlook on the climate issue, not from immediate quantifiable results.

It is said that the beauty of this project is that it is incredibly easy to plant, but it is very difficult for you to ensure that the tree you have planted stays there forever. There needs to be a newfound respect for both the flora and fauna and the passing down of such respect from one generation to the next.

This project is a star for the world to follow on a dark night. Whether Pakistan practices good politics or not, the planet does not care: it is the initiative of change and the 'want' to be better and the will to show more than what has been shown before. I understand the people, and I know the place, and for someone who has lived both an Eastern and Western life, it opens my eyes to a very realistic vision of human nature. We might come from completely different backgrounds, but the principle of protecting something fragile and beautiful is within us all! I believe in their ideals, and I know it is not easy to change behavior and cultural patterns of thought, but all is possible with time, support, and patience.

Wherever you are on this planet, whatever decision you make on a day-to-day basis, these choices are what define you. The same goes for a nation and the government that protects it. Does your country have similar projects circulating that can give you any hope?

In the process of being the first person to lay that brick, to plant that tree, to recycle that bottle—whether it is done by a family unit or an entire community—the fact is that you are initiating! You are believing in something, and the perspective of what family and friends think of you for doing it is not important. You have made a choice to be better, not just for your little bubble but for the entire world.

# 5

# BEYOND THE TREES

Pakistan has potential! It truly does. Assuming that planting 10 billion trees in the next decade will save Pakistan is naive for sure, but it will push Pakistan to make more eco-thoughtful decisions throughout. The more Pakistan practices its green mentality, the better it will be at rallying rich countries to fight for its future.

We are aware that the details of the mission are where the actual performance can happen. The accountability and credibility of policies put forward need to be a straight-up fact, put in terms that are understood and easily followed. It needs to be done as soon as humanly possible for the world to recover and for people to regain their confidence in their future and that of their children's future.

On top of caring about the image of reforestation, some large companies have stepped in to lend a supportive hand to Pakistan's efforts to start the green agenda and move economics and job creation to a greener perspective.

The trees are a brilliant idea and a wonderful message to the rest of the country, but it does not stop there. The Green Stimulus was one of those initiatives that promoted employment in green jobs sectors after the pandemic crash of 2020. PM Imran Khan agreed that improving the availability of green jobs will alleviate some of the stress from job losses and increase the awareness of the environment.

Pakistan needs to prove itself! And that is no small feat. When the government spearheaded this ambitious challenge to halt and reverse the effects of climate change on their land, they did it with a conviction that many other countries could admire. They say that nothing is more dangerous and powerful than a man with everything to lose. A county stuck 'between a rock and a hard place' will find ways to develop projects and funds that alleviate their people's stress and promote a viable future for their ecosystem and their economy. If the intention is pure, then the actions are well-founded.

Pakistan is doing something drastic about its current situation. By committing politically backed operations and a collective notion of sustainability, then change is entirely possible. What Pakistanis face now is the threat of their livelihoods by the fear and misunderstanding of the problem. Teach people about the problem because people cannot protect what they do not love. Teach them to love their soil, their air, their natural habitats, and let them see where it is all going to go if nothing changes.

And the best way to push an eco-agenda is for the communities to help with their own two hands, to promote with their own words. If a country cannot unite its people, then there is a terrible inner conflict. But Pakistan understands its people, and it understands better how to move them towards an ideal. Under the leadership of the PM, we hope for a greener, stronger, more confident country that still holds both its tradition and growth in one hand.

## THE GREEN STIMULUS

What is the Green Stimulus Project? It is an initiative to rebound from the COVID-19 losses in employment and production while at the same time creating more space for environmental awareness and mitigation. The youth of Pakistan is taking up more and more green jobs like TBTTP. This Green Stimulus is an offshoot of the program where the World Bank contributed a solid $120 million to the initiative.

Its objectives are:

- Job creation and youth empowerment
- Local ecosystem regeneration
- Upholding the Paris Accord

This is a win-win situation. Yes, the jobs created are what they call 'shovel ready,' meaning that they comply with COVID-19 distance restrictions as farming happens in open areas, but it also increases the awareness of the ecological niche and enhances the capabilities of the populous. Thus for sure, the eventual status of Pakistan as a leading country for climate change and initiative is bound to occur.

Moving to electric energy use is also a relatively common pressure point, and Pakistan could and should begin strategizing the new avenue of transportation. Using electric trucks in the large import and export of minerals and top resources is a top priority.

Amin Aslam, the PM's Special Assistant and Climate Change Advisor, stated in an official statement that "The Green Stimulus initiative has led to the generation of 87,000 green jobs for the youth rendered unemployed because of the COVID-19 pandemic" (Press Trust of India, 2021). This is a beautiful thing to read on paper, but what consequences will this initiative have on keeping the project on the high-level priority in government and a long-term solution?

The trick here (which is still being hushed out) is finding the long-term benefits of this job space. If you eliminate the prospect of creating jobs for life, what happens to that hard labor? The land that is used can be quickly returned to the farmer in two or three chops of his ax. The implementation of these reforestations needs to look at all the aspects, from indigenous species used, whether they are hardy enough to withstand the temperature, and whether they will be there long enough to see the results. It takes a human-made forest around half a century to start creating a proper carbon sink. The trees need to settle, and an ecosystem needs to develop around and within the forest, before it can start breathing. And, of course, the reasoning and communication practices with the local communities are where the magic lies.

None of these initiatives and projects, whether national or local, can make a dent if there is not a universal understanding of the problem and the solution. I would like you to think of this project as a stimulus of the green filter that Pakistan's political and governmental associations need to adopt! The phases put forward are for a set time, limit, and scale for which the country can see growth or decline. Sure, nothing happens all at once; therefore, the step-by-step actions to change the country must be remembered and kept close.

The phases of funding are what I would like to address below. Here the initiatives sprouting are brilliant and courageous, but will they be water-tight (excuse the pun)? Let us have a look.

***Phase 1: Green Jobs***

The core value of reforestation is that of protection. It is imperative that the saplings that have been planted get the chance to see their first year and grow to fulfill their purpose. Deforestation must cease immediately to allow Pakistan to take back its forested areas.

As Sara Hayat says, "Planting trees is, therefore, a welcome initiative, but can distract attention and redirect funds for other, more effective measures that can mitigate climate change and generate jobs" (Hayat, 2020). And yes, it can be obvious when the statistics of funding in eco-renewal and regeneration compared to other sectors are meager.

Apart from planting trees and restoring land, other green jobs are the project of eco-friendly infrastructure processes, where renewable energy is at its core and creating a more eco-viable environment. Some projects work towards removing cars from the roads by paving cycling lanes in urban areas. Some jobs work at removing plastics or addressing the communication towards greener lifestyles. The notion of green jobs derives from the transformation of enterprises, labor markets, and economies into a low-carbon future. Focusing on the reduction of impact, whether direct CO2 emission or health impact on workers or infrastructural damage to communities, can ultimately bring about economic sectors that do not need energy from fossil fuels and eliminate malpractice. These jobs require restoring services like water and health, food and sustainable biodiversity, and promoting activities that educate and renew communities.

The concept of promoting green jobs in sectors like energy, construction, and tourism is part of their short- and medium-term strategies. The Green Stimulus is heavily invested in training all factions of its society, from academia, civil, government, and stakeholders.

***Phase 2: Ecosystem Restoration Fund***

In 2019, at the COP25 in Madrid, Advisor to the PM, Malik Amin Aslam launched with fervent pace the Ecosystem Restoration Fund of Pakistan.

The Conference of the Parties has met up every year since 1995 to discuss strategic changes necessary to address climate change worldwide. Pakistan was at the center of it two years ago, and even though COP26 was rescheduled in 2020 for this year (no surprise there), now that it is 2021 when things seem to be getting back to normal, we still see Pakistan as a VIP position at the party.

This year, Pakistan hosted the World Environment Day on the 5th of June, 2021, and the country was extremely well received for the forward-thinking ideas, concepts, and projects that were expressed. The mention of peace, social well-being, and social justice throughout the world are admirable, but proactiveness is the missing link that many countries lack.

As we have discussed before, the mention that Pakistan is geographically and economically vulnerable is the fund's primary focus, to show the world that nothing will get better without definitive plans in place because Pakistan's proverbial time bomb is already ticking.

The fund proposes six themes to promote coordination between Pakistan's government and the global agenda for climate change:

1. Building back their precious forests one tree at a time.

2. Integrating purposeful water management solutions in all provinces.
3. The mitigation and conservation of land degradation and biodiversity in turn.
4. Moving towards the Blue Economy buzz, by also protecting fishing practices and marine wildlife.
5. Making the smart choice to evolve their eco-tourism sector.
6. Forgetting about all the petrol and diesel and working towards electric cars and hybrids.

The importance of aligning these values is critical, as Pakistan needs international support on the matter. The overall feeling that the world needs to come together financially to save what is left of this planet is the final achievement. It is sound to say that many good things can happen with encouragement and passion, but without the monetary backing for third-world countries to arrive at the starting line, then it is all a pie in the sky agenda!

***Phase 3: "Debt-for-Nature" Swap Scheme***

Here is where something is seen for the first time and is a radical and proactive initiative of its kind. The "debt-for-nature" relief says that Pakistan's debt scheme will be diminished the more efforts are seen to mitigate the challenges of climate change.

We already know that job creation is at the forefront for those who were the worst affected by the 2020 pandemic, but now the effect of those stressful months of lockdown and life cessation has jolted economies and has brought about a severe debt crisis. The monumental struggle of these developing countries to pay back their debt is where the emerging problem arises.

In Climate Home News, Chloé Farand notes, "The triple crisis of COVID-19, ballooning debt and climate change has propelled debt-for-climate schemes to the top of the international agenda" (Farand, 2021).

The Finance for Biodiversity Initiative developed this first-of-its-kind bond (absolutely brilliant idea, by the way). The acceleration of the TBTTP is where it stands, forever linking debt retirement with the targets in restoring ecosystems. These imposed targets of nature restoration will (according to the scheme) account for the writing-off of newly acquired debt or diminished return rates on loans.

This newly implemented nature performance bond will push other countries to tag along and pledge to the mission statement—the more countries that hold hands here, the more that potentially gets done.

## NATIONAL ADAPTATION FOR CLIMATE CHANGE

This plan of action is all about how the country will adapt to its surroundings. National Adaptation plans are highly critical of any government that is attempting to reverse the sands of time, so to speak. You want to put up the right walls now before the sea eats them away and the storms break down their barriers. The right way the first time! That is why the PM has pleaded for funding for Pakistan and other developing countries to be able to adapt and mitigate these effects.

Pakistan is carving out its own opportunity through the crisis, estimating a reboot of its economy in the Green Stimulus and its various phases of funding. They seem to be a lot more aware of the economic architecture that brought them through the 20th century and why it will not support them into the 21st century without proper planning.

We have spoken previously about the EbA fund, and this project falls quite neatly under the umbrella of adaptation at its finest. The 2-year project developed under the UNEP was essentially funded by the Green Climate Fund and raised a solid $2.7 million through the collaboration of government members and their ministries with private sectors and the academia surrounding climate efforts.

Pakistan is participating in various national and regional projects that develop:

- Major assistance in the technical implementation of healthy natural environments and their local use. If projects can be sustained through resilience and better governance, then all other sectors, like sanitation, water, disaster management, and agriculture, can soon follow.
- Intense investment in sharing information across the board with farmers and business owners. Their focus on education (as you have noticed prior) is absolutely critical, so they are throwing their weight toward teaching new practices and showing a smarter and cleaner way toward the future.
- Agro-business is their business! If Pakistan can get a firm grip on the future of that sector, then it can get a firm grip on many other aspects of its economy. This revolves around both changing their existing systems of crops and how they are irrigated.

## GREEN EURO BOND

Pakistan's Water and Power Development Authority (WAPDA) has been around since 1958, and its central stimulants are the operation, maintenance, and eventual upgrading of water reservoirs and power generation. The country's involvement in the expanding and eco-conscious marketplace aligned with the Green Bond Framework and its principles in 2018. Pakistan wanted to know the actual value of the assets within its borders and initiated the Natural Capital Valuation to evaluate biodiversity loss and degradation over a period of time.

The first bond is the Indus Bond, to raise $500 million in 10 years at competitive costing, and attracted many international investors to WAPDA through building dams in Diamer-Bhasha and Mohmand.

The Securities and Exchange Commission of Pakistan mentioned in a statement this year that "The issuance of green bonds will help initiate sustainable projects for combating climate change besides facilitating issuers of debt securities to diversify their course of financing. It will also provide an additional financing instrument to those investors who are conscious of financing or refinancing projects that contribute to the environment" (*SECP issues guidelines*, 2021).

The Green Bond will do a couple of things all at once, such as bringing about attractive tax incentives for investors in the bond and aiming their energy efficiency projections in future mitigation and adaptation plans. Along with Green Bonds, Blue Bonds (supporting fragile marine ecosystems) are implemented as well.

Basically, this piggy bank from international and local investors is set aside to alter all the variables that increase climate stress on a country. The funds are allocated meticulously, and budget sheets with appropriate credit balances are part of the scheme. We know money doesn't solve everything, but it sure makes it easier to see a better future when it is used for the good of humankind.

If we go beyond the trees and see the structure of a government at work, we can see many holes where funds can leak through, where policies can go to die, where things get forgotten, like an old toy under the couch. So following through with their new schemes and projects and always keeping their society and the rest of the world up to date with their step-by-step initiatives is the only way for Pakistan to build credibility. Whether they create more jobs, reduce their debt, focus on adapting economically and physically to environmental change, their real power lies in what results in the end. If they stay honest and forthright about their dreams, then the world can trust them to build this nation with integrity and honor.

I tend to advise people to remove blinders when considering their impact in any aspect of their life. We are so systematically hardwired only to see what directly affects our lives that we forget to look at what still surrounds us but does nothing for us. We become complacent and lazy with our choices and routines. The same goes for a government body throwing out plenty of schemes and projects but not realizing that just because they are out there will not automatically make them successful.

Take a moment and think about what in your life is unnecessary, even the most minor thing that could be removed, changed, or used more wisely. Are you fully conscious of your surroundings?

Remember, just like the cell is a carbon copy of your entire being, so the government needs to be a carbon copy of its people. They are the integral implementer of external influences, and if they are not aware of their superfluous actions, then neither will you be. Start by being the change you want in the world, and ask more critical questions about your singular impact physically and mentally on the environment around you.

Self-evaluation is the subject of the final chapter of our journey. Next, we will delve into what Pakistan still has to achieve and what these new achievements prove to the rest of the world.

# 6

# LESSONS FROM PAKISTAN

It is true, the goal of climate change is there, but arriving at the goal will not happen any time soon. The efforts of one country to push for change can optimize the worldview of that country, but that doesn't mean that something concrete is actually happening. We can envision a future where everything comes together, and people learn from their mistakes, but that world is still really quite far away.

Countries like Pakistan, which are in constant contact with other nations for their supply of energy demand and overall economic stability, are indeed under strain to grow without the need of bigger friends to help them along.

Then what is Pakistan standing to win in this race to eco-conscious stability?

More than anything, I believe it is a lesson, a vision, a plan, if you will, for the rest of the world's countries to adopt and follow in their own environments. If Pakistan can do it, then nothing stops the more developed countries from doing the same and much more.

Many concerns still face this growing country when creating consistent policies and procedures for maintenance, cost-cutting, and social well-being. The economic factor that plagues Pakistan will take many years of re-evaluation before it can safely show the results of its efforts.

We can say that Pakistan has the right idea but simply does not have the proper practices put in place (yet) to provide sustainable action of those ideas. This beautiful country has a long way to go, but it certainly is not afraid to tell the world how far it has come and where its aim is pointed in the climate.

The use of fossil fuels will carry on if the demand for consumption in regular energy sectors is never addressed. It is admirable to see the government start from the bottom up, using the land, the citizens, and it's own generated funds to mitigate and adapt. But it will take a lot more financial backing, incentivizing, and collective action for Pakistan to stand tall.

In this final chapter, I would like to talk about what is still on the agenda for Pakistan, where they are missing the point, and where they have gotten it just right so that other countries can look, admire, and adopt the same schemes and visions this country is applying.

By now, you must have noticed how I have spoken about my birth country with pride, hope, and trust, but also with concern, realism, and unbiased doubt. All these concepts are extremely important when you want to express your view. Being on the fence about something is small; rather than having a say for both sides, understanding where they fix and where they harm is a much more mature, worldly way of looking at ideas, and I hope I have shown you how to do the same.

## STILL A LOT LEFT TO DO

Recognition: This word is everything when it comes to the humble fight for a better and cleaner Pakistan and a safer and sounder world. We can all understand that it is important to fix corrupt systems. A misfire in the system can be seen in that Pakistan's international climate financing is much lower than those of other vulnerable countries. This is seen as a fundamental shortfall from the Ministry of Climate Change, where they have no technical knowledge on how to integrate climate funding methods. Those proposals that could bring in a 'bankable' investment with nongovernment organizations and private companies often get rejected.

Here are some major facets of Pakistani affairs that need to be addressed by the government within the next five years, with an iron fist in a velvet glove approach:

***Corruption in Public and Private Affairs***

Even with the Prevention of Corruption Act and the National Accountability Bureau Ordinance implemented along with other smaller anti-corruption legislations, there is still an alarmingly high rate of corruption and resistance to change that has brought about inefficient and absent control. Pakistan's lack of accountability in social health, education, judiciary, and police are visible. The same goes for the price of renewables in the Pakistani market and the precious resource of water. Gangs and criminal institutions are everywhere in major cities, and the police have not got the firmest grip on the problem. The fact that the poorer citizens are left in the dust so the more affluent population can survive is outrageous. The divide is increasing year by year, where communities are being stripped of their value by the technologies and modern practices that can seem harmless at first but deteriorate values and common goods in the long run.

***Loopholes in Governance and Policies***

This is no more evident than in the speech government officials make on building a more renewable-based country when still, in black and white, they are one of the highest users of coal and other fossil fuels in the world. It is estimated that a third of Pakistan's energy is sourced from coal (around 19%) through CPEC funded plants (Isaad, 2021). This is aggravated by the recent Pakistan and China initiative to build six more coal plants in the next decade. This seems like one step forward and three steps back, but obviously, phasing out coal will have drastic effects on the economy. Therefore, if the government and power producers would revise their contracts and re-evaluate where they can reduce and where they need more time for planning. That would be a more transparent technique to incentivize the people rather than just fluffing it up with grand ideas.

***Pollution From Radical Urbanization***

"Despite being economic engines, cities in Pakistan suffer from stresses like climate change, haphazard and unregulated expansion, housing shortage, and a lack of basic civic amenities" (Abdul & Yu, 2020). This rides on the back of the climate issue hovering over major cities (mainly in the southern regions of Pakistan) that are struck by heat waves, CO2 clogging, waste mismanagement, and potable water stress. Cities like Karachi, Sukkur, Badin, and Islamabad show not just the impact of heat, flood, rain, and space deprivation. But socio-spatial segregation, displacement, and degradation of the surroundings by agriculture, industry, and the general blindness to consequence. Turning these urban areas into ecological havens is on the top of the government's list, and creating the Green Pakistan movement was at its core. Incompetent engineering, destructive beautification projects, and the ever-present greedy developers will not subside until some aggressive policies are put down and hard!

### *Deforestation Blind Spots*

The final fact here is that the local residents cannot simply stop using the free and easy-to-get wood fuel to sustain their lives without proper access to electricity or natural gas. The government is responsible for evaluating the underlying reasons for the people's need to cut down forests. This is the more delicate nature of the project that needs to be addressed soon, even if the zero-tolerance policy on large-scale illegal logging is underway.

### *Unregulated Water Resources*

This runs deep within Pakistani governance. Their water resources are in peril. Without the necessary steps taken to conserve, re-route, and manage water resources in a more modern and sustainable way, Pakistan will be suffering heavily from dehydration by 2050.

### *Educating the Masses*

The future of the world lies in its youth, which I think we can all agree on. In the next 50 years, what we do will pave the way for new generations to see behind the veil of politics and economics and strive for real hand-in-hand change in social welfare, integration, inclusion, and overall world peace. Poor health through disease, hunger, thirst, and negative environments all boil down to the strategies of those in power. Yes, I have spoken about how the government can do only so much and that the country's population needs to come together. Still, before that is possible, the government needs to throw everything it has into giving the people a *reason* to care. And the only way that happens is when education is at the head of the strategy.

The religious implications of Islamic countries could help with this education effort. Muslim nations can remind people from Islamic sources of the value of preserving the environment. Education is only useful when applied to all aspects of the environment, politics, social justice, and religion. If the government can boast about transparency and modern shifts in thought to the rest of the world, then reflecting that through the education of men and, most importantly, women in its country is the paradigm shift that will turn the tides.

So far, we have touched on the many actions that government and the public are doing to change this narrative for future generations, but let us have a look at them once more in summary:

- Afforestation and reforestation
- Infrastructures, such as dam and rainwater catchment sites
- Water management and hygiene
- Using electric or hybrid vehicles in major transport systems
- Switching to renewables like solar, wind, hydro, and so forth
- Acknowledging different climate approaches for different regions
- Forgetting about reactiveness and focusing on proactiveness
- Push the political agenda through climate action in real, tangible scenarios

The need for long-term strategies in natural resource management is where the magic lies. Where governments are inclined to seek short-term solutions to the problems they face to complete within a tenure, they need to instead focus on projecting change over long-term temporal spaces.

Nature will not work harder to grow or recover. It will go precisely when it intends to. Thus the lack of total human control over the matter can bring insecurity and mismanaged and sudden actions. These strategies might be pretty to look at, but they need to be a whole lot more useful to both an individual country and other countries looking in.

The vast projects to increase mangrove cover over the last 30 years is estimated to be around 300%. This is wonderful news, but the concern comes from more indigenous species being wiped out by overfishing, camel grazing, and deforestation.

Numbers can look good if you only show the world what you think they need to see. That is where Pakistan is falling short, and it is understandable, as they need to promote their adaptability. But it is vital that no wool is thrown over anybody's eyes during this process. All statistics and evaluations need to be thorough and transparent, with no cutting corners and procedures to look apart artificially.

Agricultural water is not taxed, and the canal system in which water is distributed to the urban areas is undervalued and underpriced. Theft and other black market operations are a norm now in places like Karachi and other big cities in Pakistan. This is expected when for so long, this country could not care because it didn't have the means to. Thank goodness the tides have changed!

## WHAT WE CAN LEARN FROM PAKISTAN

The World Bank has been a solid partner in Pakistan's initiative to reverse or better combat the sands of time. The world has stood up and cheered about how well the country is doing and how they have leap-frogged the traditional system and touched upon some successful critical aspects.

But instead of just praise, the bigger idea is that the world follows the story of Pakistan more closely and recognizes where it will succeed and fail. By keeping a close eye on the implications of change, the financial barriers of a developing country that is still reeling from independence, and its unfortunate location in geographical terms, other developing countries may find encouragement to do the same.

If Pakistan can be the first to show the way, then that is a success story all on its own.

### *Leapfrogging It*

As described by Yayboke et al., "According to traditional theories to development, the path to prosperity for emerging economies is to follow in the tracks of developed nations. The theory goes that if countries wish to become wealthier, they should simply commit to following the same series of steps that allow developed economies to prosper" (Yayboke et al., 2020). But this theory in recent years has been overlooked and put aside for something far more innovative: the leapfrog development method.

This new concept of leverage is growing in popularity and seeing some potentiality in accelerating growth. This means that instead of going the old route of doing as the others do, which nine times out of ten ends in the middle-income trap (countries getting stuck at a certain level of necessity due to the glass ceiling effect). Instead, the country bypasses steps and goes straight to the source. Modern technology and alternative renewable resources are aimed at in the first phase, giving the government a compelling goal of long-term development without the unnecessary 'rights of passage' that often do nothing for real improvement.

Low-income countries can take advantage of new technologies and knowledge-based economy structures to pass over low-value manufacturing that plagues their development. The most common example is that of the digital revolution, where mobile phones precede the outdated progress of landline infrastructure, which in itself is a backward concept. Developing countries like Cambodia and Uganda are also implementing this strategy in the digital space and focusing on science and technology to uplift and upgrade their social-economic status.

Pakistan has taken great strides to overcome these processes and clear a space for such innovation in its own right. Focusing on research and development, education, infrastructure development, and intellectual property protection is the fundamental base of this strategy. The government must understand that implementing the enthusiasm correctly is based on proper infrastructure, policymaking, and carefully addressing developmental gaps.

Both with their opportunity to harness incredible amounts of solar energy, thanks to the arid climate, and with their wind and hydropower infrastructure plans requiring investment in the utility-scale of renewable energy power plants, the nation could significantly reduce the cost of electricity by not importing that 60% coal, oil, and gas.

The Union of Concerned Scientists mentioned in early 2020 that the opportunity for Pakistan to shift away from fossil fuel dependency and rely on renewables is there; it is just waiting to be realized. They state that richer countries will need to "provide the finances that countries like Pakistan need to make this transition happen more quickly. It is not fair that Pakistan takes on the cost of doing this all on its own" (Hadid, 2020), especially if Western countries are the biggest emission culprits.

***Plastic Activism***

Pakistan has started envisioning some bright new ideas to integrate its nation and future generations into the eco-conscious pathway. They have begun to play with the concept of filling the infrastructure gap with more innovative resources, and one of them is through the World Without Plastic Program developed this year. It sees the country re-evaluate how they build roads and what their building material comprises. The outcome is a tri-factor win by mixing recycled PET plastic (8 tons of plastic bottles and bags) and standard construction material. First, the roads are 15 times more durable with plastic introduced. They are cost-effective as less concrete mix is supplied. They are sustainable due to their innovative use of a pollutant in construction through community and company recycling.

Although this is not a novel idea and has been used in various other countries around the world, this is undoubtedly the first time a public-private partnership has come about in Pakistan towards this smart, sustainable project. The re-carpeting of Ataturk Avenue in Islamabad is a collective effort between Coca-Cola Pakistan–Afghanistan and the Capital Development Authority. The Vice President of Coca-Cola Pakistan–Afghanistan, Fahad Ashraf, recently said, "The idea provides a breakthrough solution to bring back plastic into the productive economy. And we also want to focus on building a community around the idea and the innovation itself" (Qureshi, 2021).

Along with the World Without Plastic movement, we can also appreciate the Say No to Plastic campaign's efforts that stand against plastic use and littering. It was initiated in 2014 when the first-ever climate change public hearing was to grace Pakistan's Upper House Parliament. This saw hundreds of students demonstrate in the capital of Islamabad toward informed recycling and plastic use, asking residents to reconsider what they buy and what they can re-use, with the attempt to eliminate plastic shopping bags as well.

### *Smart Financing*

Financial reliance is a deep dark pit that many countries have never been able to get out of. Pakistan itself is stuck in one of those poverty-ridden economies reliant on exterior assistance to keep the government running. We see financing from international support and from within the country and its sectors. But we are surely not so naive as to think it came pro bono.

But now, Pakistan's government and policymakers are implementing intelligent, practical, and viable solutions to increase international funding by promoting transparent and resilient planning into their policies. By adapting to the future of a low-carbon economy and building resilience for their most vulnerable people and areas, the government can increase co-funding and allow international organizations to commit to the strategy. Pakistan is showing the world that:

- Strengthening domestic institutional capabilities allow for a better assessment of how they can manage international funds in the future.
- They translated their message for climate change into its economic costs while prioritizing direct adaptations versus policymaking.

- They are focusing on the local context of the space and adjusting the requirement according to the needs of each province.
- They are using the high-levels of government (the PM) as voice and fist through the challenge of change.

The Financing Framework for Resilience Growth states that they need to determine the adaptation gap, track public expenditure, estimate economic loss and damage, fill the adaptation gap, and identify future financing scenarios (Malik, 2018).

Through confusion and uncertainty of the 2020 pandemic, Pakistan stood out from the rest and showed up to the climate mitigation and adaptation party in a big way! When everyone was hiding away at home, this country took up its shovels. It continued to plant trees, demonstrate awareness of the issue, and commit to (albeit not always achieving) and monitor the change in economic and social welfare.

You can see the effects of climate change on your own country, no matter where in the world you live. This is Earth's own pandemic, its own disease that is spreading and corroding the planet. We have created this monster with our own hands, so it will have to end with our own hands.

Carbon capturing is the way forward according to some of the most influential countries in the world, like China, Australia, and Japan. Yes, cutting CO2 costs is vital, but many industries would never survive the switch to renewable energy in the timeframe determined, meaning that they will still have to rely on fossil fuels for their power. A viable solution to this issue is called 'carbon capture and storage,' where emissions are not released into the atmosphere but rather deposited into a capturing facility and sent to the North Sea for storage under the sea bed. This is said to be a very safe method of reducing the emission cost on the planet but could also bring more costs in the process. This could indeed be a safe option forward, but the number one priority is reducing to a 2 degree Celsius and 1.5 degree Celsius global temperature at reachable goals.

As COP26 is to be held in Glasgow in November 2021, the Intergovernmental Panel on Climate Change that evaluates and assesses each country's report results said they could not accept these new allegations. Some countries alleged that specific stipulations within the assessment requirements should be eliminated, as they uphold something far greater than words. Nations like Japan, Australia, and Saudi Arabia came together to lobby for altering and removing specific vital reports. For example, "The need for urgent and accelerated mitigation actions at all scales" (Rowlatt & Gerken,

2021) was mentioned not to be as vital as the vital component stipulates.

Therefore, there should be no deviations from the major points established by the UN for climate change, and pointing fingers at newer methods to combat global warming is not sufficient without proper factual statistics and cost-effectiveness.

If individuals, corporations, and governments come together to better the state of their country, with a sprinkle of positivity, perseverance, and pragmatism, each person can make that change. I certainly do not want to end this book with some sweet, callow tone that perceives a happy ending for everyone, but I would like to convey a sense of hope.

The question is: Do you have hope for the world? Is it enough for you that we just try to stand against our own selfish needs, or do we need to dig deeper for a real change?

What I would like to know is whether you have read through my work and comprehended the fickle nature of change. It can almost come to its epitome without action, without reason, and this is the change that will kill the majority of life on the planet. The change we need is the active, choice-worthy type of change.

I like to consider climate change the same way I consider the concept of marriage. You would most likely want to fight for the right change to occur because if you left the relationship to run its course without intervention, the change would still be there, just in the form of separation, not reconciliation. We need to *reconcile* with planet earth, which takes a lot of courage, faith, and addressing complex issues over a period of time.

And there you have it: time! It seems to be the only thing we have and the only thing we are racing against. Climate resilience will not happen today, tomorrow, or next week. It will happen precisely when the people change.

# CONCLUSION

The final judgment of humanity and civilization depends on not turning away and making it someone else's problem but on actively including every nation, religion, and demographic to uphold better environmental values and more robust long-term strategies. Saving the future requires enormous amounts of recognition, responsibility, collaboration, and accountability from all facets of society.

Pakistan is showing enthusiasm and drive, so much that their commendable words can become a reality if the people in charge relinquish greed and corruption and replace it with affirmative action and sustainable funding. Mahmood Akhtar Cheema, a representative for the IUCN, says, "Pakistan is ahead of many when it comes to planning, but the execution of the plan remains the challenge" (Zaman, 2021). And yes, that is the ultimate speed bump for recovery and growth, but it is a beginning!

Pakistan is under severe threat but they are prepared to make a stand and hopefully become one of the world's first developing countries to adapt and mitigate Earth's dire situation and grow their economy and international presence along the way too. We can only really realize the threat posed when we perceive it first-hand. Unfortunately, media and politics have altered those critical points to render a different message to the populous in the Western world.

But remember, as one always must, that it is never too late to try. Sure, we are later than hoped to catch on, but we are young beings, and if we carry on propagating a message of 'the best place to get a helping hand is at the end of your own arm' theory, then we can truly begin to see the change from the roots up.

In this book, I have introduced quite a few topics, and it might have felt a tad overwhelming at times, but that, to me, is the feeling of change. We touched on various institutions, policies, promises, triumphs, strategies, and collaborations that have been spoken about and acted on by Pakistan's government and its people. We talked about where Pakistan is standing right now and why they are sitting in dangerously hot water. We spoke about the effects their past, and present actions have had on their environment and economy. We explored what the public and private sectors are doing to address each facet of the problem by adding green thinking into water consumption and conservation, land use, urbanization, job creation, education, transport, and international affairs.

We dived into the nitty-gritty of PM Imran Khan's grand scheme to plant a courageous amount of trees and all those companies and individuals who have been pulled into the project. The sequestration of CO2 is a primary target that the PTI seems to push for with very positive outcomes. We can see where the light shines bright for the country through praise and criticism, but we can also see where the shade of poverty, corruption, and misinformation can hinder its time in the spotlight.

I have constantly spoken about how intuitive the new government's strategies are and how invested they are in their plan, from funding out of their own wallets to coordinating with international associations to relieve debt smartly and thoroughly exploiting newer options for energy production and greener construction. By consistently including the general public to assist and promote the message, the efforts come across as living, breathing concepts that are propagated not for the benefit of the one but the many.

And finally, we got around to talking about what the country still has to accomplish in terms of infrastructure renewal and maintenance, unemployment, corruption, inclusion, education, and thorough policy makeovers. By acknowledging what they have done and how far they have come, Pakistan's grand plans can be a guiding light for other nations around the world. Pakistan is one of those countries that is beginning to reach in that gap and pull out cognitive and creative resources to fight for quantifiable change while also finding better ways to jump to the next phase of growth.

What I hope you have absorbed is the *idea* of the message. If the message is strong, with a solid foundation, a solid backing, and the ability to be monitored and evaluated, then people will listen.

Like when writing this book, for example, the best way to catch attention is by dropping the critical points: the real, desperate, and devastatingly ignored issues that destroy lives. These get mulled over in the reader's head, start making sense, and from doubt comes sadness and anger. Then, we show the solution and the change that is possible, giving the reader a sense of hope and encouragement. This method of message delivery is one of the purest physiological manipulators out there; it does not hurt the reader; it simply cries for urgency and brings an inevitable sense of external pressure.

If the people do not have the right access to world news, to modern mechanics of information and communication, then it is indeed up to the government and leaders of that country to push the message further. They need to make sure to use correct context, whether scientific or religious, to guide their citizens into joining the climate change team and understanding the rules of the game.

And so, Pakistan's vision is brought across with urgency and hope. How the people of its land are told the story of its sickness in its flooded rivers, its melting glaciers, and its shrinking forests, the story of its smog-covered cities, its water-starved citizens, and overall health crises will be told affect how people react. But the tale does not have to end in tragedy. No, this is a tale of rebirth. Pakistan shows us where we can plant the first tree, where we can reduce our usage, how we can educate our youth, and how they plan to create a green economy. This is the trick, and it works as long as that story is held true with faith and trust between people and their government.

And thanks to the efforts of international subsidiary companies and the cooperation of other South Asian nations, Pakistan can see a sustainable and green future; even if it is far, that future is there. They just need to stay on the right path.

So, what have you gotten out of this book? Has it impacted you in any way, and are you better for it? Of course, this whole book has been about revision. Throughout, I have questioned the ethics, morals, institutions, and overall governance of a country, but have you done the same for your own? And if you have, what were your findings?

I know you have just absorbed a ton of information so far. You have read how people are coming together to change, but have you been wondering where you can even begin to take the first step towards promoting a cleaner and greener version of your own world?

Let me give you some tips:

- Contacting a representative in your region, area, the province is the first step. Find your team and bring everyone you know who cares about it as much as you (if they don't, make them care!) together to approach the government's button pushers and ask, demand, discuss change, and support.
- Reach out in a formal letter or email to your representatives and leaders and express clearly and technically (with relevant $CO_2$ and

pollution statistics, please!) why you believe there needs to be more action.

- While you wait for actions to be processed at higher levels, you need to find ways to reduce your own waste behaviors and carbon footprint. You do this by carpooling, using public transport, recycling, using less plastic, walking more, never throwing away food or clothes, growing your own vegetable garden, volunteering at local support centers and small-scale farming groups, and using fewer heating and cooling systems in your household. Sure, converting to renewable green energy products (eco-fridge, oven, microwave, shower, toilet, electric cars) is good but mighty expensive. So do what you can now, understand what you contribute to your emission cycle (you can get an app that roughly calculates your daily CO2 emission), and find innovative ways to keep living your good life, but this time sustainably!

Dear reader, please remember that you are only human and that your single contribution (whether it be a few weeks or the rest of your life) will not change the world. The world is too big, and it is too complicated and too intertwined. Humble and simplistic solutions are great but small. What will change the world is your voice, your actions to bring more people together, and guidance for others to follow a cleaner and greener way of life. Power is in numbers, and by advocating and making it a priority in your community and in your electoral sessions, you might be able to see the message sink in even deeper.

We are told multiple messages about the change and evolution of a species, that we are at the top of our game, and that the world has never been happier. There has never been a more 'content' time in human history than now. We live in a world that is obsessed with having enough. The whole market screams this at you, where we have more food in the dumps than in our bellies. We all need to be more aware of this, and we need to change that narrative one step at a time. We need to simply understand that the concern is not the supply-demand aspect but the consumption and conservation aspect.

Thank you again for picking up this book. It means everything to me that there are people who still believe and care about the bigger picture, the sad picture, the more ominous picture. You have seen something new, learned about a country that reflects the same but very different principles than your own, and is just trying to get out of this mess in one piece, like everyone else.

If you think this book has done anything for you, please do not forget to drop a review and express your own personal thoughts about the words written and the message given. Thank you again, from the bottom of my heart and from the people of my homeland.

# REFERENCES

*2020 Karachi floods*. (2021, October 27). Wikipedia. https://en.wikipedia.org/wiki/2020_Karachi_floods

Abdul, L., & Yu, T. (2020, December 14). *Resilient urbanization: A systematic review on urban discourse in Pakistan.* Urban Science. https://www.mdpi.com/2413-8851/4/4/76

Alam, A. R. (2019, August 20). *PTI's one year: Fixing how we deal with the climate crisis*. Dawn. https://www.dawn.com/news/1500606

Ali, K. (2021, May 20). *New policy to focus on promoting electric vehicles*. Dawn. https://www.dawn.com/news/1624564

Anwar, N. H. (2017, September 20). *Cities, climate change and Pakistan's extended urbanisation.* Dawn. https://www.dawn.com/news/1358905

Aslam, M. A. (2021, February 12). *How Pakistan is aiming for a green recovery from the pandemic.* World Economic Forum. https://www.weforum.org/agenda/2021/02/pakistan-green-recovery/

Baloch, S. M. (2018, June 7). *Water crisis: Why is Pakistan running dry*? Deutsche Welle. https://www.dw.com/en/water-crisis-why-is-pakistan-running-dry/a-44110280

Clean Green Pakistan. (2021). *Clean Green Pakistan.* Ministry of Climate Change. https://cleangreen.gov.pk/eng

Dawn Staff Reporter. (2021, January 27). *10 Billion Trees Project praised in UK parliament.* Dawn. https://www.dawn.com/news/1603768

Ebrahim, Z. T. (2020, December 17). *Scrapping imported coal projects, Pakistan fails to let go of local lignite.* Eco-Business. https://www.eco-business.com/news/scrapping-imported-coal-projects-pakistan-fails-to-let-go-of-local-lignite/

Eckstein, D., Künzel, V., & Schäfer, L. (2021). *Global Climate Risk Index 2021.* https://reliefweb.int/sites/

reliefweb.int/files/resources/Global%20Climate%20Risk%20Index%202021_1_0.pdf

*Effects of climate change*. (2021). World Wildlife Fund. https://www.worldwildlife.org/threats/effects-of-climate-change#:~:text=More%20frequent%20and%20intense%20 drought

Farand, C. (2021, April 16). *Pakistan explores debt-for-nature scheme to accelerate its 10 billion tree tsunami*. Climate Home News. https://www.climatechangenews.com/2021/04/16/pakistan-explores-debt-nature-scheme-accelerate-10-billion-tree-tsunami/

Finnigan, C. (2019, April 5). *Fighting climate change in Pakistan: One school at a time*. London School of Economics and Political Science. https://blogs.lse.ac.uk/southasia/2019/04/05/fighting-climate-change-in-pakistan-one-school-at-a-time/

*Forestry in Pakistan*. (2019, October 24). Wikipedia. https://en.wikipedia.org/wiki/Forestry_in_Pakistan

Gauhar, T. (2020, November 19). *Fixing Pakistan's power sector—A roadmap*. The News International. https://www.thenews.com.pk/print/745917-fixing-pakistan-s-power-sector-a-roadmap

*Global fund for ecosystem-based adaptation*. (2020, February 3). IUCN. https://www.iucn.org/theme/

ecosystem-management/our-work/ecosystem-based-approaches-climate-change-adaptation/global-fund-ecosystem-based-adaptation

Hadid, D. (2020, November 29). *With glaciers melting and temps soaring, Pakistan pursues big action on climate change*. NPR. https://www.npr.org/2020/09/29/916878679/with-glaciers-melting-and-temps-soaring-pakistan-pursues-big-action-on-climate-c

Harris, B. (2020, January 22). *5 quotes from Pakistan's Prime Minister at Davos*. World Economic Forum. https://www.weforum.org/agenda/2020/01/5-key-quotes-from-pakistans-prime-minister-imran-khan-at-davos/

Hayat, S. (2020, June 2). *Opinion: Pakistan needs to go beyond tree planting to thrive after Covid-19*. The Third Pole. https://www.thethirdpole.net/en/climate/opinion-pakistan-needs-to-go-beyond-tree-planting-to-thrive-after-covid-19/

Hersher, R. (2018, July 27). *Hot weather spells trouble for nuclear power plants in Europe*. NPR. https://www.npr.org/2018/07/27/632988813/hot-weather-spells-trouble-for-nuclear-power-plants

Hess, L. (2021, June 8). *Pakistan is on its way to planting a 10 Billion Tree Tsunami. Here's how*. Landscape News. https://news.globallandscapesforum.org/52934/

pakistan-is-on-its-way-to-planting-10-billion-tree-tsunami-heres-how/

Hu, X., Imran, M., Wu, M., Moon, H. C., & Liu, X. (2020). Alternative to oil and gas: Review of economic benefits and potential of wind power in Pakistan. *Mathematical Problems in Engineering,* 2020(8884228), 1–16. https://doi.org/10.1155/2020/8884228

Husain, K. (2021, March 21). *Pakistan's surplus power generation capacity has come at a price*. Scroll.in. https://scroll.in/article/989919/pakistans-surplus-power-generation-capacity-has-come-at-a-price

Hussain, M. (2019, April 17). *Pakistan's battle against climate change*. ReliefWeb. https://reliefweb.int/report/pakistan/pakistan-s-battle-against-climate-change

Hutt, R. (2018, July 2). *Pakistan has planted over a billion trees*. World Economic Forum. https://www.weforum.org/agenda/2018/07/pakistan-s-billion-tree-tsunami-is-astonishing/

Isaad, H. (2021, January 18). *Opinion: Is Pakistan really phasing out coal?* The Third Pole. https://www.thethirdpole.net/en/energy/opinion-is-pakistan-really-phasing-out-coal/

Jamieson, A. (2018, August 5). New leader of terror-torn Pakistan plans to plant 10 billion trees. *NBC News.*

https://www.nbcnews.com/news/world/pakistan-plans-plant-10-billion-trees-fight-climate-change-n897216

Junaidi, I. (2019, December 31). *128,000 deaths occur in Pakistan due to climate change every year, Senate body told.* Dawn. https://www.dawn.com/news/1525374

Kanwal, S., Ding, X., Sajjad, M., & Abbas, S. (2019, December 18). *Three decades of coastal changes in Sindh, Pakistan (1989–2018): A geospatial assessment*. ResearchGate. https://www.researchgate.net/publication/338009311_Three_Decades_of_Coastal_Changes_in_Sindh_Pakistan_1989-2018_A_Geospatial_Assessment

*Key terms you need to know to understand climate change.* (2015, October 27). The Climate Reality Project. https://www.climaterealityproject.org/blog/key-terms-you-need-understand-climate-change

Khadim, M. (2017, November 2). *List of top 15 health issues in Pakistan*. Transparent Hands. https://www.transparenthands.org/list-of-top-15-health-issues-in-pakistan/

Khan, H. N. (2020, November 6). *Climate change adaptations for sustainable agriculture in Pakistan*. [Video] YouTube, TEDx Talks. https://www.youtube.com/watch?v=sxUQTL7yMGw&t=97s

Khan, M. I., Hussain, S. K., Saad, H., Rukh, G., Ahmed, M. M., & Ahmad, I. (2017). *Third party monitoring of billion trees afforestation project in Khyber Pakhtunkhwa Phase-II*. WWF Pakistan. https://d2ouvy59p0dg6k.cloudfront.net/downloads/btap_monitoring_report_phase_ii.pdf

Khan, R. S. (2020, September 22). *"Freak events": Karachi floods hint at shifting monsoon*. Reuters. https://www.reuters.com/article/us-pakistan-monsoon-climatechange-idUSKCN26D1MS

Khan, S. (2020, September 8). *Pakistan: Climate change, environmental problems put government in a bind.* Deutsche Welle. https://www.dw.com/en/climate-change-puts-pakistan-in-a-bind/a-54849791

Khan, S. (2021a, April 1). *Is Pakistan prepared to deal with climate migration*? Deutsche Welle. https://www.dw.com/en/pakistan-climate-migration-crisis/a-56126512

Khan, S. (2021b, September 2). *Pakistan: Environmentalists slam '10 billion trees' project* Deutsche Welle. https://www.dw.com/en/pakistan-environmentalists-slam-10-billion-trees-project/a-59062200

Kugelman, M. (2017, March 9). *Pakistan's unheralded fight against climate change*. New Security Beat. https://

www.newsecuritybeat.org/2017/03/pakistans-unheralded-fight-climate-change/

Latif, A. (2021, June 4). *Pakistan's fight to reverse climate change gains recognition*. Anadolu Agency. https://www.aa.com.tr/en/asia-pacific/pakistan-s-fight-to-reverse-climate-change-gains-recognition/2263517

Lázaro, C. (2020). *Building back better after COVID-19: Pakistan's Protected Areas Initiative*. Protected Planet. https://www.protectedplanet.net/news-and-stories/building-back-better-after-covid-19-pakistans-protected-areas-initiative

Malik, M. M. (2018, July 24). *A financing framework for resilient growth—What it means for Pakistan*. Climate and Development Knowledge Network. https://cdkn.org/2018/07/a-financing-framework-for-resilient-growth-what-it-means-for-pakistan/?loclang=en_gb

Malik, S., Awan, H., & Khan, N. (2012). Mapping vulnerability to climate change and its repercussions on human health in Pakistan. *Globalization and Health*, 8(1), 31. https://doi.org/10.1186/1744-8603-8-31

McGrath, M. (2020, November 4). *Climate change: US formally withdraws from Paris Agreement*. BBC News. https://www.bbc.com/news/science-environment-54797743

Mughal, R. (2020, January 14). *Pakistan extreme weather death toll at 83.* VOA. https://www.voanews.com/a/south-central-asia_pakistan-extreme-weather-death-toll-83/6182566.html

Mukhtar, I. (2020, September 30). *With water scarce, Pakistan helps farmers grow more with less.* Reuters. https://www.reuters.com/article/us-pakistan-water-climatechange-farming-idUSKBN26L0BN

*Nationally Determined Contributions (NDCs).* (2021). United Nations Climate Change. https://unfccc.int/process-and-meetings/the-paris-agreement/nationally-determined-contributions-ndcs/nationally-determined-contributions-ndcs

*Olive cultivation to make Pakistan self-sufficient in edible oil.* (2021, April 26). The Nation. https://nation.com.pk/26-Apr-2021/olive-cultivation-to-make-pakistan-self-sufficient-in-edible-oil

*Pakistan ramps up protected areas.* (2021, May 28). UN Environment Programme. https://www.unep.org/news-and-stories/story/pakistan-ramps-protected-areas

*Pakistan to develop national adaptation plan for climate change.* (2021, March 25). UNEP. https://www.unep.org/gan/news/press-release/pakistan-develop-national-adaptation-plan-climate-change

*Pakistan's Billion Tree Tsunami restores 350,000 hectares of forests and degraded land to surpass Bonn Challenge commitment.* (2017, August 11). IUCN. https://www.iucn.org/news/forests/201708/pakistan%E2%80%99s-billion-tree-tsunami-restores-350000-hectares-forests-and-degraded-land-surpass-bonn-challenge-commitment

*Pakistan's Ten Billion Tree Tsunami.* (2021, June 2). UN Environment Programme. https://www.unep.org/news-and-stories/story/pakistans-ten-billion-tree-tsunami

Pakistan's climate change. (2018, January 10). *Pakistan's climate change financing framework.* UNDP Government of Climate Finance for Asia-Pacific. https://www.youtube.com/watch?v=yimC9b8SvSE

PBS. (2021, September 25). *Pakistani Prime Minister Imran Khan speaks at 2021 U.N. General Assembly.* [Video] YouTube. https://www.youtube.com/watch?v=bHoWDBu-e1Y

Press Trust of India. (2021, April 2). *World Bank to provide $120 mn to Pakistan for its Green Stimulus initiative.* Press Trust of India. https://www.business-standard.com/article/international/world-bank-to-provide-120-mn-to-pakistan-for-its-green-stimulus-initiative-121040200008_1.html

Qasim, M. (2020, September 18). *Pakistan leapfrogging to a green energy future*. East Asia Forum. https://www.eastasiaforum.org/2020/09/18/pakistan-leapfrogging-to-a-green-energy-future/

Qureshi, Z. (2021, September 24). *Pakistan's first plastic road project kicks off under "World Without Plastic" programme*. Gulf News. https://gulfnews.com/world/asia/pakistan/pakistans-first-plastic-road-project-kicks-off-under-world-without-plastic-programme-1.82488716

Recharge Pakistan. (2020, December 11). *"Recharge Pakistan" launched to enhance climate resilience and water security*. The Nation. https://nation.com.pk/11-Dec-2020/recharge-pakistan-launched-to-enhance-climate-resilience-and-water-security

*Report: China emissions exceed all developed nations combined*. (2021, May 7). BBC News. https://www.bbc.com/news/world-asia-57018837

Roth, K. (2021). *World Report 2021: Rights trends in Pakistan*. Human Rights Watch. https://www.hrw.org/world-report/2021/country-chapters/pakistan

Rowlatt, J., & Gerken, T. (2021, October 21). *COP26: Document leak reveals nations lobbying to change key climate report*. BBC News. https://www.bbc.com/news/science-environment-58982445

Saeed, H. (2020, September 22). *IUCN, FAO and WWF Pakistan join hands with Ministry of Climate Change to monitor the largest afforestation programme in Pakistan.* IUCN. https://www.iucn.org/news/pakistan/202009/iucn-fao-and-wwf-pakistan-join-hands-ministry-climate-change-monitor-largest-afforestation-programme-pakistan

Salam, A. (2018, 24 July). A heatwave has killed 65 people in Pakistan's Karachi. *USA Today*. https://www.usatoday.com/story/news/world/2018/07/24/pakistan-one-worlds-leading-victims-global-warming/809509002/

*SECP issues guidelines for green bonds issuance in Pakistan.* (2021, June 19). The News. https://www.thenews.com.pk/print/851615-secp-issues-guidelines-for-green-bonds-issuance-in-pakistan

Shams, S. (2019, November 23). *Why religious narratives are crucial to tackling climate change*. Deutsche Welle. https://www.dw.com/en/why-religious-narratives-are-crucial-to-tackling-climate-change/a-51384374

Stults, S. (2015, October). *Poem about environmental preservation, Warned*. Family Friend Poems. https://www.familyfriendpoems.com/poem/warned

*The climate threat: Opportunities and challenges in Pakistan's agriculture.* (2021, July 29). Center for

Strategic and International Studies. https://www.csis.org/events/climate-threat-opportunities-and-challenges-pakistans-agriculture

The Info Light. (2021, September 23). *UK Prime Minister Boris Johnson praises PM Imran Khan in United National.* [Video] YouTube. https://www.youtube.com/watch?v=Qk2gwFVnOyI

The News International. (2021, July 8). *Govt to launch Olive Trees Tsunami Project*. The News International. https://www.thenews.com.pk/print/861136-govt-to-launch-olive-trees-tsunami-project

*Topography of Pakistan*. (2019, December 22). Wikipedia. https://en.wikipedia.org/wiki/Topography_of_Pakistan

*UK PM hails Billion Tree Project.* (2021, September 23). Pakistan Today. https://www.pakistantoday.com.pk/2021/09/23/uk-pm-hails-billion-tree-project-at-un/

Van Bronkhorst, B., & Bhandari, P. (2021). *Pakistan climate risk country profile*. World Bank Group and Asian Development Bank. https://climateknowledgeportal.worldbank.org/sites/default/files/2021-05/15078-WB_Pakistan%20Country%20Profile-WEB.pdf

Vener, J., & Jamshaid, S. H. (2021, August 11). *Pakistani young people have their say on climate*. United Nations

Development Programme. https://www.undp.org/blog/pakistani-young-people-have-their-say-climate

Webmaster. (2019, December 11). *Launch of Pakistan's ecosystem restoration fund at COP25, Madrid, Spain.* National Disaster & Risk Management Fund. https://www.ndrmf.pk/pressreleases/launch-of-pakistans-ecosystem-restoration-fund-at-cop25-madrid-spain/

*What you can do to fight climate change.* (2019) World Wildlife Fund. https://www.worldwildlife.org/pages/what-you-can-do-to-fight-climate-change

WION. (2021, July 6). *Gravitas: What's keeping Pakistan in the dark?* [Video]. YouTube. https://www.youtube.com/watch?v=kqJnHNMeoxE

World Bank. (2020, June 29). *Pakistan undertakes reforms to bolster fiscal resilience and COVID-19 recovery.* The World Bank. https://www.worldbank.org/en/news/press-release/2020/06/29/pakistan-undertakes-reforms-to-bolster-fiscal-resilience-and-covid-19-recovery

WorldEnvironmentDay. (2021, June 4). *Financing arrangements for Ecosystem.* [Video] YouTube. https://www.youtube.com/watch?v=DWpxhpcVlho

World Environment Day. (2021, June 5). *World Environment Day: Pakistan and its "tree tsunami" project in the spotlight.* The Star. https://www.thestar.com.my/lifestyle/living/2021/06/05/world-environment-day-pakistan-and-its-039tree-tsunami039-project-in-the-spotlight

Yayboke, E., Carter, W. A., & Crumpler, W. (2020, April 10). *The need for a leapfrog strategy.* CSIS. https://www.csis.org/analysis/need-leapfrog-strategy

Zaidi, R. (2021, February 22). *Pakistan to host World Environment Day 2021.* UN Environment Programme. https://www.unep.org/sw/node/28961

Zaman, Q. (2021, July 9). *Pakistan eyes world's first "nature performance bond."* The Third Pole. https://www.thethirdpole.net/en/nature/nature-performance-bond-pakistan/

Made in United States
North Haven, CT
04 November 2026

21769187R00106